Data Modeling for Azure Data Services

Implement professional data design and structures in Azure

Peter ter Braake

BIRMINGHAM—MUMBAI

Data Modeling for Azure Data Services

Copyright © 2021 Packt Publishing

All rights reserved. No part of this book may be reproduced, stored in a retrieval system, or transmitted in any form or by any means, without the prior written permission of the publisher, except in the case of brief quotations embedded in critical articles or reviews.

Every effort has been made in the preparation of this book to ensure the accuracy of the information presented. However, the information contained in this book is sold without warranty, either express or implied. Neither the author(s), nor Packt Publishing or its dealers and distributors, will be held liable for any damages caused or alleged to have been caused directly or indirectly by this book.

Packt Publishing has endeavored to provide trademark information about all of the companies and products mentioned in this book by the appropriate use of capitals. However, Packt Publishing cannot guarantee the accuracy of this information.

Group Product Manager: Kunal Parikh

Publishing Product Manager: Aditi Gour

Senior Editor: Roshan Kumar

Content Development Editors: Sean Lobo, Priyanka Soam

Technical Editor: Arjun Varma

Copy Editor: Safis Editing

Language Support Editor: Safis Editing

Project Coordinator: Aparna Ravikumar Nair

Proofreader: Safis Editing

Indexer: Rekha Nair

Production Designer: Alishon Mendonca

First published: July 2021

Production reference: 1240621

Published by Packt Publishing Ltd.

Livery Place

35 Livery Street

Birmingham

B3 2PB, UK.

ISBN: 978-1-80107-734-7

www.packt.com

Contributors

About the author

Peter ter Braake started working as a developer in 1996 after studying physics in Utrecht, the Netherlands. Databases and business intelligence piqued his interest the most, leading to him specializing in SQL Server and its business intelligence components. He has worked with Power BI from the tool's very beginnings. Peter started working as an independent contractor in 2008. This has enabled him to divide his time between teaching data-related classes, consulting with customers, and writing articles and books.

About the reviewers

Marcel Alsdorf has been working for 4 years as a cloud solution architect at Microsoft. In his role, he consults companies on cloud computing fundamentals, the construction of modern data warehouse pipelines, event systems, and machine learning applications and tooling. In addition, he lectures on big data fundamentals at a German university. Previously, he worked as an FPGA engineer for the LHC project at CERN and as a software engineer in the banking industry.

I would like to thank Christoph Körner for suggesting me as a reviewer and the team at Packt for giving me the opportunity to help Peter ter Braake create a good starting point for people in terms of understanding the fundamentals of data modeling in Azure.

Alexey Bokov is an experienced Azure architect and has been a Microsoft technical evangelist since 2011 – he works closely with Microsoft top-tier customers all around the world to develop applications based on the Azure cloud platform. Building cloud-based applications in the most challenging scenarios is his passion, as well as helping the development community to upskill and learn new things by working hands-on and hacking. He has been a long-time contributor, as a co-author and reviewer, to many Azure books and is an occasional speaker at Kubernetes events.

I'd like to thank my family – my beautiful wife, Yana, and my amazing son, Kostya, who have supported my efforts to help the author and publisher of this book.

Table of Contents

Section 2 – Analytics with a Data Lake and Data Warehouse

7

Dimensional Modeling

Table of Contents

3

Normalizing Data

4

Provisioning and Implementing an Azure SQL DB

5

Designing a NoSQL Database

6

Provisioning and Implementing an Azure Cosmos DB Database

Section 2 – Analytics with a Data Lake and Data Warehouse

7

Dimensional Modeling

8

Provisioning and Implementing an Azure Synapse SQL Pool

9

Data Vault Modeling

10
Designing and Implementing a Data Lake Using Azure Storage

Section 3 – ETL with Azure Data Factory

11
Implementing ETL Using Azure Data Factory

Other Books You May Enjoy

Index

Preface

Databases play an important role in almost all the applications that we use. The database has a direct impact on the performance and scalability of the application it supports. That makes choosing the right type of database to use and designing that database correctly a vital part of all development since scalability and performance depend on a well-chosen design. With databases hosted in Azure, the design may also have a direct impact on the costs of the database. The first part of this book teaches you when to use a relational database (Azure SQL Database) and when a NoSQL database (Cosmos DB) is the better option. You will also learn how to design that database and, finally, how to implement the chosen design.

All the data gathered by applications can be used for Business Intelligence (BI). A crucial part of BI is creating a central repository for your data to build your BI solution on. This can be a data lake with data marts, or it may be a data warehouse. In the second part of the book, you will learn to design data warehouses according to the theory of dimensional modeling or by designing a data vault. You will also learn how to design a data lake. You will then learn how to apply what you have learned by creating a data lake in Azure Storage and creating data marts using Azure Synapse Analytics.

The book ends with a chapter on Azure Data Factory. Data Factory is used to get data from the source databases you created in part one and store that data in the data platforms you implemented in part two.

After reading this book, you have a solid understanding of data modeling and of how to implement a database schema in Azure. This will help you to build scalable and cost-effective data solutions in Azure.

Who this book is for

This book is for application developers, BI developers, and data engineers who build databases and data solutions in the Microsoft Azure cloud. This book assumes that you have basic knowledge of Azure and a basic understanding of what a database is and what it is for. The book does not assume any specific development or database knowledge or experience. However, you should have an affinity for working with data. Some T-SQL experience will also be advantageous.

This book will also help data analysts gain an understanding of how databases are set up and why they are created in the way they are. This understanding will help data analysts to get the most out of the different data stores that they may need to use.

What this book covers

Chapter 1, Introduction to Databases, explains what a database is and what it is used for. It also explains the common concepts of databases.

Chapter 2, Entity Analysis, explains how to create and read **Entity Relationship Diagrams (ERD)** from a top-down perspective.

Chapter 3, Normalizing Data, teaches you how to create a normalized database design and when to use this design technique.

Chapter 4, Provisioning and Implementing an Azure SQL Database, shows hands-on how to create and implement a normalized design in Azure using Microsoft's PaaS offering – Azure SQL Database.

Chapter 5, Designing a NoSQL Database, explains when to choose Cosmos DB over a relational database. It also teaches you how to design different types of NoSQL databases.

Chapter 6, Provisioning and Implementing an Azure Cosmos DB, shows hands-on how to create and implement a Document database in Azure using Microsoft's NoSQL database, Cosmos DB.

Chapter 7, Dimensional Modeling, teaches you how to create a star schema database design according to the rules of dimensional modeling.

Chapter 8, Provisioning and Implementing an Azure Synapse SQL Pool, shows hands-on how to create and implement a star schema database using an Azure Synapse dedicated SQL pool.

Chapter 9, Data Vault Modeling, explains when to use a data vault and how to design a data warehouse using the data vault modeling technique.

Chapter 10, Designing and Implementing a Data Lake Using Azure Storage, discusses when implementing a data lake is a better option than creating a data warehouse. It also shows how to implement a data lake using Azure Storage.

Chapter 11, Implementing ETL Using Azure Data Factory, shows how to create pipelines to automate the process of getting data out of production databases and into data lakes and data marts.

To get the most out of this book

Since this book implements all database services in Azure, a computer with a modern browser and an Azure subscription with permission to create resources in Azure are required:

Software/hardware covered in the book	Operating system requirements
Azure Data Services: SQL Database, Cosmos DB, Storage, Synapse Analytics	Windows, Linux, MacOS

If you are using the digital version of this book, we advise you to type the code yourself or access the code from the book's GitHub repository (a link is available in the next section). Doing so will help you avoid any potential errors related to the copying and pasting of code.

Download the example code files

You can download the example code files for this book from GitHub at `https://github.com/PacktPublishing/Data-modelling-for-Azure-Data-Services`. If there's an update to the code, it will be updated in the GitHub repository.

We also have other code bundles from our rich catalog of books and videos available at `https://github.com/PacktPublishing/`. Check them out!

Download the color images

We also provide a PDF file that has color images of the screenshots and diagrams used in this book. You can download it here: `https://static.packt-cdn.com/downloads/9781801077347_ColorImages.pdf`.

Conventions used

There are a number of text conventions used throughout this book.

`Code in text`: Indicates code words in text, database table names, folder names, filenames, file extensions, pathnames, dummy URLs, user input, and Twitter handles. Here is an example: "Column names such as PatientID, Patient_FirstName, and PostalCode would already render it more readable."

A block of code is set as follows:

```
CREATE TABLE [dbo].[OrderDetail]
(
OrderID      INT NOT NULL,
ProductID    INT NOT NULL,
UnitPrice    MONEY NOT NULL,
Quantity     SMALLINT NOT NULL,
Discount     NUMERIC(5, 4) NOT NULL
);
```

GOWhen we wish to draw your attention to a particular part of a code block, the relevant lines or items are set in bold:

```
[default]
exten => s,1,Dial(Zap/1|30)
exten => s,2,Voicemail(u100)
exten => s,102,Voicemail(b100)
exten => i,1,Voicemail(s0)
```

Any command-line input or output is written as follows:

```
$ mkdir css
$ cd css
```

Bold: Indicates a new term, an important word, or words that you see on screen. For instance, words in menus or dialog boxes appear in **bold**. Here is an example: "Click, in Azure Data Studio, on the **New Query** button to open a new query file."

> **Tips or important notes**
> Appear like this.

Get in touch

Feedback from our readers is always welcome.

General feedback: If you have questions about any aspect of this book, email us at customercare@packtpub.com and mention the book title in the subject of your message.

Errata: Although we have taken every care to ensure the accuracy of our content, mistakes do happen. If you have found a mistake in this book, we would be grateful if you would report this to us. Please visit www.packtpub.com/support/errata and fill in the form.

Piracy: If you come across any illegal copies of our works in any form on the internet, we would be grateful if you would provide us with the location address or website name. Please contact us at copyright@packt.com with a link to the material.

If you are interested in becoming an author: If there is a topic that you have expertise in and you are interested in either writing or contributing to a book, please visit authors.packtpub.com.

Share Your Thoughts

Once you've read *Data Modeling for Azure Data Services*, we'd love to hear your thoughts! Scan the QR code below to go straight to the Amazon review page for this book and share your feedback.

https://packt.link/r/1-801-07734-7

Your review is important to us and the tech community and will help us make sure we're delivering excellent quality content.

Section 1 – Operational/OLTP Databases

Databases are used in everyday applications such as webshops, CRM systems, and financial systems. These are our **Line Of Business** (**LOB**) applications. In this section, we will learn how to choose between a SQL database and a Cosmos DB database. We will also learn how to design a database for optimal performance and scalability.

This section comprises the following chapters:

- *Chapter 1, Introduction to Databases*
- *Chapter 2, Entity Analysis*
- *Chapter 3, Normalizing Data*
- *Chapter 4, Provisioning and Implementing an Azure SQL DB*
- *Chapter 5, Designing a NoSQL Database*
- *Chapter 6, Provisioning and Implementing an Azure Cosmos DB Database*

1
Introduction to Databases

Data has become increasingly important over the last few years. Almost all applications use data, whether the application is a **Customer Relationship Management (CRM)** system at work or a social media app on your phone. All that data is stored in databases. Since the 1980s, almost all those databases have been relational databases. Nowadays, with the advent of big data, there are different ways to store and process huge amounts of data. Some of them can be classified as so-called NoSQL databases. NoSQL stands for "not only" SQL. This means that we are seeing other types of databases emerge and being used alongside relational databases. NoSQL databases are important in the area of big data. The "SQL" in NoSQL stands for **Structured Query Language**. This is the programming language of relational databases and has become the "equivalent" of relational databases.

In this chapter, you will learn the basics of databases. A lot of the theory discussed in this chapter stems from relational databases, although the majority is applicable to other database systems as well.

We will discuss the following topics in this chapter:

- Overview of relational databases
- Introduction to Structured Query Language

- Impact of intended usage patterns on database design

- Understanding relational theory

- Keys

- Types of workload

Overview of relational databases

Databases hadn't yet been invented when we first started programming computer applications. All data had to be stored in files. Oftentimes, those files were simple **comma-separated value** files (**CSV** files). An example of a CSV file can be seen in the following screenshot:

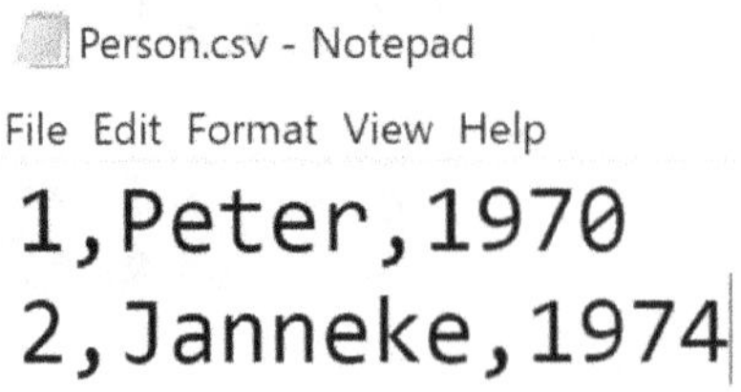

Person.csv - Notepad

File Edit Format View Help

```
1,Peter,1970
2,Janneke,1974
```

Figure 1.1 – Person.csv

As you can see, it is just some data without anything else.

Files

Using files to store data for use in applications entailed a number of issues. After trying file formats other than CSV files, developers started using databases instead of plain files. Plain files or flat files are files with just data stored in them. *Figure 1.1* is an example of a flat file. Let's look into the issues that using flat files posed.

From the header of the screenshot in *Figure 1.1*, it is clear that the file is called `Person.csv`. We may infer that the data in the files represents persons. However, it is not clear whether those people are patients, customers, employees, or even someone completely different. Furthermore, you cannot ascertain that extra information from the file or its content.

Drawbacks

The use of these types of flat files to store data comes with three drawbacks:

- You cannot infer from the file itself what the data is about.

- It is not flexible from a programming perspective and is bad for performance when working with the data.

- It is (almost) impossible for multiple persons to work with flat files simultaneously.

We will now examine each of these drawbacks in turn.

Drawback 1 – You cannot infer from the file itself what the data is about

It is clear from looking at the screenshot that each line has two commas, meaning that there are three columns per row. The second column very likely holds a first name. This is a reasonable assumption based on our knowledge of names, although you may require a knowledge of Dutch names to make this assumption. The third column is more difficult to guess. It could be the year of birth of the person in question, but it could also be a postal code or perhaps a monthly salary.

The file only stores the actual data and not the metadata. It may be that you can guess what the values mean, but you cannot infer it from the file itself. Metadata is the data describing the data. Column names are an example of metadata. Column names such as PatientID, Patient_FirstName, and PostalCode would already render it more readable. That is why we often add those column names as a first row in flat files.

> **Note**
> *Metadata is data that describes the "actual" data.*

There is even more to ascertain regarding this data. You cannot perform calculations with postal codes, such as adding up two postal codes (it may be that you can, but it doesn't make any sense). A postal code is an alphanumeric code that you cannot perform computations with. When the last column in *Figure 1.1* is a salary and not a postal code, you do want (and need) to be able to perform calculations on this column, for instance, to calculate an annual salary from the monthly salaries. In this case, the column would have been numerical. In other words, it would be beneficial to know a column's data type. Generally speaking, data can be numerical, alphanumerical (text), or dates. Nowadays, of course, there are a lot of variations, such as binary data for pictures.

With data stored in flat files, the data itself and the metadata are stored separately.

Today, we have overcome some of these issues by not using flat files but storing data as XML or as JSON files. Both file types allow you to store metadata with the actual data in the file itself. In the (recent) past, this was too expensive to do. Only recently has storage become cheap enough and compute power plentiful enough to work with text files by storing data and metadata in the way that JSON does.

Drawback 2 – It is not flexible from a programming perspective and is bad for performance when working with the data

It gets nastier when we start using (old-fashioned) program code to work with the data. Suppose you need to know the postal code of the person called Janneke. Your code would now look something like this:

1. Read a line.

2. Read the second column.

3. If the value you read equals *Janneke*, then return the third column.

4. Repeat lines 1 to 3 until there are no more lines in the file.

With only two lines in the file, this is pretty fast. This code will become more problematic, however, when a file contains many, many rows. It will become really slow.

It gets even worse when someone changes the file structure. Suppose we add a new column, storing the patient's family name between the second and third columns. The code we just wrote will break because it assumes that the postal code is the third column. However, following the change, it is the fourth column. Retrieving the postal code should be independent of which column it actually is.

Drawback 3 – It is (almost) impossible for multiple persons to work with flat files simultaneously

In most applications, there will be multiple users working with the data simultaneously. What if your webshop could only have one visitor at a time? A database should make it easy for multiple people or processes to work with the same data at the same time. In the case of relational databases, this is part of their core. A relational database has what is known as the **ACID** properties to cater to multi-user workloads. You will learn more about the ACID properties in *Chapter 5, Designing a NoSQL Database*. Without a database system, whether relational or not, multiple users working with the same data would not be impossible, but you will get consistency issues if you don't implement complex logic to prevent inconsistencies.

If you always process all the data in a flat file as a whole, and you do that, for instance, during the night, flat files are fine to work with, as we will see in *Chapter 10*, *Designing and Implementing a Data Lake Using Azure Storage*. However, if you need to work with individual pieces of information from within a flat file in real time, you will not be able to do that in an acceptable manner.

At first, smart workarounds were invented to make working with flat files easier and more efficient. There were files such as ISAM files and VSAM files. It is beyond the scope of this book to go into these different file types. More interesting for us is the fact that the problems described in this paragraph led to the introduction of database management systems (DBMSes).

Relational databases

A database is a self-describing collection of related data with the aim of providing information to people and applications.

The first database appeared in the 1960s. These databases were hierarchical databases. A little later, network databases were introduced, but neither type of database offered the flexibility to work with (large amounts of) data in more complex organizations with multiple users.

In the early 1970s, E.F. Codd, an English mathematician working for IBM, came up with a theory of how to create relational databases. He based his theory on mathematical set theory. This theory describes sets of elements that are potentially really large in a few simple rules (that will be covered later in the chapter). Codd realized that mathematical set theory could not only be applied to something abstract such as *all even numbers*, but also to real live collections such as *all our customers*. This rendered set-based theory useable in relation to the data we were working with and the data we needed to store in databases.

The name *relational database* stems from the fact that data is stored in tables. For example, take a set of numbers {1, 2, 3, 4}. Then, imagine a second set, for instance, a set of names {Peter, Janneke, Jari, Mats}. We could combine these two sets in a table, as shown in *Figure 1.2*:

PatientID (int)	PatientName (nvarchar(50))
1	Peter
2	Janneke
3	Jari
4	Mats

Figure 1.2 – A table of patients

We started with independent sets of values. We created a relation between the two sets by combining them into a table. Making a table with rows is like saying the values *1* and *Peter* belong together, just as *2* and *Janneke* do. This makes the sets no longer independent. A relationship now exists between the values in one column and the values in another column. In other words, the table is the relation between the **PatientID** set and the **PatientName** set. *Relation* here is another word for *Table*.

> **Note**
> Relational databases store data in tables.

We see something more in *Figure 1.2*. The first column is called **PatientID**. The column header also specifies that the data type of the values in this column is **int**. This means that this column can only store whole numbers. The second column is of the **nvarchar(50)** type, specifying that it stores alphanumeric values (text) with a maximum length of 50 characters. This metadata is part of the table itself. The data and the metadata are now a whole instead of separately stored pieces of information.

> **Note**
> In a database, data and metadata are combined in a single structure.

A relational database is normally more than just one table. In real life, a relational database can consist of thousands of tables. According to the definition of a database, it is a collection of related data. This means that the tables have relationships with one another. Since *relationship* sounds a lot more like *relational* than *table* does, a lot of people came to believe that a relational database got its name from related tables. However, as stated previously, storing data in tables is what makes a database relational.

Relational Database Management System

A **Relational Database Management System** (**RDBMS**) is a piece of software that allows you to create and manage databases that adhere to Codd's theory. That turns an RDBMS into an application that allows you to create tables and then store data in those tables. The "management" part is all the extra "services" you get from an RDBMS, such as securing your data so that only authorized people can work with the data. An RDBMS allows you to work with data, from creating tables to storing and managing the data and its accessibility. Examples of well-known RDBMS systems include Microsoft SQL Server, Oracle, IBM DB2, MySQL, and MariaDB.

> **Note**
>
> An RDBMS is a database product that follows Codd's rules of the relational model, allowing you to work with and manage all your data.

We previously referred to a couple of problems that we encountered in the past when using CSV files. One was the lack of metadata. Relational databases rectify that problem. The question that remains to be answered is how relational databases offer the flexibility and performance needed that CSV files couldn't offer. To do that, we first need to introduce the SQL language.

Introduction to Structured Query Language

Structured Query Language (**SQL**) is the language of all relational databases. You use SQL to read data from the database and to manipulate existing data. Creating a database or creating tables within the database and securing your data is also done using SQL. Database developers might use a tool to graphically create databases or tables, so you don't need to write SQL code yourself. The tool will generate the SQL code for you because that is all the actual database engine understands.

Different categories of SQL

SQL consists of three main categories:

- **DCL – Data Control Language**
- **DDL – Data Definition Language**
- **DML – Data Manipulation Language**

Data Control Language

DCL is the part of SQL that helps to secure the data. DCL comprises three statements:

- GRANT
- DENY
- REVOKE

Even though securing data is very important in any data solution that you build, DCL is outside the scope of this book. You should check out the TutorialRide website to learn more about this: https://www.tutorialride.com/dbms/sql-data-control-language-dcl.htm.

Data Definition Language

With DDL statements, you create databases themselves and objects within databases. As with DCL, there are three statements:

- CREATE
- ALTER
- DROP

With CREATE TABLE, you can make (create) a new table. Whenever the table structure needs to change, for instance, you want to add a column to an existing table, you use UPDATE TABLE. With DROP TABLE, you completely remove a table and all its content. You will learn about these statements in *Chapter 4, Provisioning and Implementing an Azure SQL DB.*

Data Manipulation Language

DML is the part of SQL that is used for working with the actual data stored in the tables in the database. DML has four statements:

- SELECT
- INSERT
- UPDATE
- DELETE

The SELECT statement lets you read data from tables. Some people believe that SELECT should be a category of its own: *DQL* or *Data Query Language.* The other three statements are self-explanatory: INSERT adds new rows to an existing table, UPDATE changes the values of already existing rows in a table, and DELETE removes rows from a table.

This book is not about SQL. There are a lot of tutorials on the internet on SQL. If you need to familiarize yourself with T-SQL (the dialect of SQL Server that we will use throughout this book), I strongly recommend the books of Itzik Ben-Gan. SQL is also used a lot in NoSQL databases and is still the basis for every data professional.

Understanding the database schema

With relational databases, the first step is to create a table using CREATE TABLE. While creating the table, you specify the name of the new table. You also specify all the columns of that table by adding a column name and the data type of the column. This means you start by creating the metadata. All the metadata combined is referred to as the **database schema**.

Often, people merely mean the table structure, the tables, and their relationships when they use the term *schema*. In SQL Server, there is even an object called a schema that helps in establishing a good security strategy for your database.

> **Note**
> The schema of a database refers to all tables and their relationships.

Once you have created tables, you can start loading data into the table using the INSERT statement. With each row you enter, you provide values for all the columns in the table. The values you enter should follow the rules defined in the table definition. With what we have learned so far, this means that the values should be of the correct data type. You cannot enter text in a numerical column. You will see shortly that there are further constraints. The process of creating a table first and then entering data means that whatever data we add (for example, INSERT) has to adhere to that structure.

After you have entered data into the database, you can start working with the data. With the SELECT statement, you can read data from the database. The database uses the existing metadata while retrieving data from the database. For instance, you could write the following SELECT statement:

```
SELECT
     PostalCode
FROM
     Persons
WHERE
     Name = 'Janneke';
```

Notice that this statement is the same example as described in the section about files. In the preceding snippet, we read the PostalCode column from the Persons table for the row that has 'Janneke' as the value in the Name column. The database uses the metadata, in this case, the table name and column names, to retrieve the data and, where possible, to optimize the query. By using metadata, it doesn't matter whether the PostalCode field is the second or the third column. Using metadata makes querying the data more flexible.

In addition to the flexibility we gained by using a table over a flat file, there is no step such as *repeat this for each row*, as we saw in the section on files. A table in a relational database implicitly works with all rows. That is called *working set-based*. In relational databases, you don't work with individual rows but always with sets. Depending on the filters you provide (the WHERE clause of the SQL statement), your set might contain just one row or it might contain multiple rows, or it can even be an empty set with no rows at all. The database engine can optimize the query for fast query response times. The database engine uses its knowledge of the metadata. It can also take advantage of extra structures that you may define in the database, such as indexes. You will learn about indexes in *Chapter 4, Provisioning and Implementing an Azure SQL DB.*

> **Note**
> Database systems take advantage of all the metadata defined in the schema to optimize queries in order to obtain good query performance even in the case of large datasets.

Now that we have learned why (relational) databases work better for storing data than CSV files, let's look at the different use cases of databases and the impact this has on database design.

Impact of intended usage patterns on database design

The simplest database consists of just one table. However, as has already been mentioned in the section on files, most (if not all) databases have multiple tables, up to tens of thousands of tables, and sometimes even more. The most important step in creating a new database is to decide which tables you need in order to store your data efficiently for its intended use and what relationships those tables have with one another. Data modeling is the process of deciding on that table structure and thereby deciding on the schema.

> **Note**
> Data modeling is the process of deciding which data structures to use for storing data. With regard to relational databases, this translates into deciding on the table structure (schema) to use.

There are different concepts for modeling data structures. We can, for example, normalize databases, invented by E.F. Codd. E.F. Codd translated mathematical set theory into formal steps to reach an optimal schema. Normalizing a database involves following these formal steps. It is often said that a database is in the third normal form, meaning the steps have been applied to the data until the third step. You will learn how to normalize a database in *Chapter 3, Normalizing Data*.

In *Chapter 7, Dimensional Modeling*, and *Chapter 9, Data Vault Modeling*, you will learn about alternative data modeling techniques. These alternative techniques lead to different database schemas. Query performance and query complexity depend heavily on the schema you choose. In Azure, this may translate directly into how expensive your data solution is going to be. If you need more processing power to execute a query, you might have to provision a higher performance tier of the service you use, which will cost more money. With a different intended use, meaning you use different queries, a different schema may be beneficial.

A database supporting a web shop should be able to quickly retrieve all information about a specific product. In addition, it should be easy and quick to create a new shopping basket as well as new invoices. This type of database is best designed by using the technique of normalization. A data warehouse should be able to quickly retrieve vast amounts of data and aggregate that data. A report based on a data warehouse should, for instance, show all the sales data of an entire product group over the last 12 months. Where a webshop often retrieves single rows from the database, a report often retrieves millions of rows. A dimensionally modeled database will be better in retrieving millions of rows than a normalized database.

In *Chapter 5, Designing a NoSQL Database*, we will look at NoSQL databases. NoSQL databases have their own rules depending on the type of database you use.

Whatever database you use, the similarities will be bigger than the differences from a modeling perspective. This is especially true for all types of relational databases. A database is always a collection of data, a set of data, that can be described by set theory. The next section will describe how the theory behind databases has its foundations in mathematical set theory.

Understanding relational theory

Let's look in a bit more depth into relational databases. They form a big part of everyday processes working with data. They also constitute the majority of this book, and what you will learn about relational theory will also help in using non-relational databases.

A relational database stores data in tables according to Codd's rules of the relational model.

Relational theory has two important pillars:

- Elements of a set are not ordered.
- All elements in a set are unique.

Pillar 1 – Elements of a set are not ordered

The first pillar from set-based theory is both neat and, on occasion, a bit troublesome.

I once had to visit a customer of the company I worked for because he had an issue. Whenever he added a new order to his database manually, he checked to see whether everything was OK by opening the **Orders** table and scrolling down to the last row. Whenever he failed to find his new order, he would enter it again. At the end of the month, he always had too many orders in his database!

In this example, my customer assumed that the row entered last would automatically be the last row in the table or the row showed last on his screen. He assumed opening a table in his application would retrieve the rows (orders) in the same order as in which they were entered. Both assumptions are incorrect (although, in theory, the latter depends on the application in which you open a table and not on the database).

A database doesn't store rows in any specific order. This provides both flexibility and write performance. However, a database administrator or developer could create indexes on tables. By using indexes, we can speed up data retrieval (query) times, in the same way an index in a book makes it quicker to find a specific topic. Certain types of indexes store the data ordered, giving the database the option to use the sorting order to quickly find rows that we are looking for. That index might be based on the customer's last name and not on the date the order was inserted in the database. Furthermore, a Database Administrator (DBA) might decide to replace the index with another one. In that case, the data is stored in a different order than it was previously. You will learn about indexes in the various chapters that zoom in on specific Azure services, such as Azure SQL Database.

Without indexes, rows are stored without a specific sorting order. With indexes, that might change depending on the index you use. But if you query the database, it is not guaranteed that the database will actually use the index. When writing a query, you can explicitly specify the sorting orders for all the rows you want returned. That sorting order is now guaranteed. If you don't ask for a specific sorting order, you will get the rows back in a random order. That order is determined by whether the database decided to use an index and additional factors, such as fragmentation in your database.

When using a relational database, the lack of order is sometimes confusing. People almost always think with some sort of ordering in mind, such as the last, best, and biggest. As long as you remember to explicitly specify a sorting order when querying a database, you will be fine. If you don't need your rows to be in a specific order, don't request it. The sorting of (large) datasets is always expensive.

> **Tip**
> When you require rows to be in a specific order, you need to specify that sorting order explicitly in your query.

Pillar 2 – All elements in a set are unique

In mathematics, a *set* is a collection of unique elements. For example, the set of uneven numbers consists of all uneven numbers. The number *3* is part of that set. It does not appear twice in the set. Each uneven number is in the set only once. This is called the **axiom of unicity**.

When applying mathematical set-based theory to everyday live databases, unicity comes in very handy. A table storing customers is the set of all customers. In a business-to-business scenario, a business is either a customer or it is not a customer. When it is a customer, we store the details in the **Customer** table, otherwise, we don't. We do not store every potential customer and we preferably do not store the same customer twice.

Unicity demands that each element of a set is unique.

Suppose a company is stored twice in your database. That will lead to a myriad of problems. To start with, simply counting the number of rows to determine how many customers you have will lead to the wrong result. A simple report showing sales by customer will also no longer be correct, nor will average sales by customer. The list goes on. You want customers to be unique within your database. As a different example, have a look at *Figure 1.3*. Can you tell how many patients live in Amsterdam?

PatientID (int)	PatientName (nvarchar(50))	City (nvarchar(50))
1	Peter	Amsterdam
1	Peter	Amsterdam
2	Janneke	Den Bosch
3	Jari	Tilburg
4	Mats	Utrecht

Figure 1.3 – Patients with duplicate rows

Apart from functional reasons for keeping rows unique in a table, there are technical reasons as well. Suppose you have a table like the one in *Figure 1.3*. You decide that the first two rows describe the same patient, so you want to remove one of the rows from the table. You cannot do that with SQL. You can delete the row(s) where **PatientID** equals **1**. The database will delete both rows. However, removing **Peter** or removing rows with **Amsterdam** also leads to both rows being removed. To delete a single row, you must be able to uniquely identify that row. In the example shown, there is no characteristic (column) separating row 1 from row 2. You cannot delete one without deleting the other at the same time.

Some RDBMS systems keep track of rows by adding a unique identifier under the covers. In that case, you can retrieve that value and delete one row using this value. Microsoft SQL Server is not one of those systems. You need to make sure rows are unique (or live with the consequences).

To make sure each row in a table is unique, you (the database administrator) should check for each newly inserted row that does not already exist. The simplest way you can do that is by comparing the new row column by column to see whether you already have a row with identical values for each column. This is easy enough, but really time-consuming. With 10 million rows in a table and 40 columns, this means you need to perform 400 million comparisons. You need to do that each time you enter a new row or change an existing row. You could utilize more optimized search algorithms but they will still not make this a fast operation.

> **Note**
>
> To enforce the uniqueness of rows, we use keys. A key is a column, or a couple of columns together, in which we store unique values.

A **key** is a column, or a combination of columns, that stores unique values by the nature of the data stored in the column. For instance, each car has a unique license plate. When you create a table that stores all the vehicles within your company, each vehicle has its own unique license plate. You can identify a specific car by its license plate, independent of the values of other columns within the table.

The context in which we store data can be important when it comes to columns storing unique values. Consider a table where we store a row of information each time a driver has damage to their car. Some people will have damage more often than others. Each time we store the details about the damage, we keep track of which vehicle has damage by storing the license plate. In the **Vehicle** table, the license plate will be unique, while in the **Damage** table, it will not be unique. You can see this in *Figure 1.4*. The **Vehicle** table holds one row per car. Each row describes a unique car with its unique license plate. The **Damage** table shows the claims of damage done to the car. Some cars have never had any claims made in relation to them, while others can be found in the table multiple times. This means that the same license plate is no longer unique:

Vehicle			
License plate	Make	Model	Owner
NL-277-J	Ford	Focus	Peter
6-TKD-30	Peugeot	107	Janneke
14-MTB-5	Lamborghini	Hurcan	Mats

Damage		
License plate	Date	Claim
6-TKD-30	1-jan-2020	1000,-
6-TKD-30	12-feb-2021	800,-
14-MTB-5	25-dec-2020	9000,-
14-MTB-5	7-mar-2021	5850,-

Figure 1.4 – Vehicles and damage

Let's look closer now into how defining keys helps with keeping rows unique in your database.

Keys

Databases do not check rows for uniqueness automatically. Checking rows column by column is too expensive. The database developer needs to define a key. After you define a key, the database will use this to check for uniqueness. You need to define a key per table. You need to decide for each table whether or not you require (want) a key. You then need to decide which column to use as the key.

Types of keys

There are different types of keys. Let's have a look at some definitions for keys.

Candidate keys

Candidate keys are all columns within a table that logically store unique values. A table may have zero, one, or more columns that, by their nature, should store unique values in the context of the table they are in. Each column that adheres to the definition of a key is a candidate key. That makes the license plate in the **Vehicle** table of *Figure 1.4* a candidate key. Other examples of candidate keys can be a **Social Security Number** (**SSN**) in a **Persons** table or an ISBN number in a **Book** table. An SSN or an ISBN uniquely identifies a person or a book, just like a license plate uniquely identifies a vehicle.

A key can be made up of several columns. In the Netherlands, the combination of postal code and house number is unique for an address. Websites can use this by automatically filling in a street name and city after you enter just your postal code and house number.

We can distinguish between two types of candidate keys – logical, or business, keys and technical, or surrogate, keys.

Logical keys

A **logical key** or **business key** is a key that stores values (information) with real meaning that you need to store, regardless of whether it can function as a key. No matter how you design your database, irrespective of whether you plan to use keys, the business key will be part of the data you are going to store. For example, all companies are obliged to store the SSN of their employees. Since the SSN column is part of your table, you can use it as a key.

In Data Vault theory (*Chapter 9, Data Vault Modeling*), business keys play a central role. Knowing your data (and the processes that use the data) becomes crucial. The business key is the characteristic that people use in their everyday lives to refer to objects. A course about data modeling is referred to (searched and found by) its name. The name is the business key. Names can be unique, but very often are not. That makes them dangerous to use as keys. Sometimes, however, they work OK.

Technical/surrogate keys

In some cases, there is not a single candidate key to be found. Adding an extra column and using unique values for it then becomes an option. That way, you are adding (creating) a candidate key. This is also an option when you do have candidate keys to start with. This type of key is called a **technical key** or **surrogate key**. Its contents are (normally) meaningless but unique.

Most database systems have features to easily create surrogate keys. For instance, a sequence is an object in the database that allows you to generate unique numbers. Each time you ask the sequence for a value, it provides you with one that is greater than the last time you asked. Each value is guaranteed to be used a maximum of one time. With a sequence, you can create efficient surrogate keys. Within SQL Server, you can create a column in a table with an extra property, identity. An identity column in SQL Server gets an automatic unique value each time a row is inserted in that table. The result is the same as with a sequence. Identity is easier in terms of its use. Both methods provide you with a candidate key.

A good surrogate key is meaningless. That means that you cannot infer anything from its value. In France, an automobile's license plate shows you where the car comes from. The code has some intrinsic meaning. In the Netherlands, a license plate has no meaning other than the fact that it is a unique code. If you keep track of how license plates are generated, you will be able to tell how old a car is. But that is basic information from another table. You cannot tell where in the Netherlands the car was registered or what type of car it is from its license plate.

Meaningless values are more flexible in terms of their use. You don't get into any sort of trouble by misinterpreting key values. You also don't encounter issues if, for some reason, the meaning of the values changes over time.

A surrogate key should be meaningless (most of the time).

Everybody knows surrogate keys. Most of the products you buy have a product number. You have an account number for your bank account. You probably have an insurance policy number, a customer number, and so on. All are surrogate keys in a database to make rows unique in that particular database.

In databases, there are two kinds of keys that you will define that play a crucial role in the database: primary keys and foreign keys. Let's have a look at them, starting with the primary key.

Primary keys

All that has been said hitherto in relation to keys is an introduction to **primary keys**. Since checking the uniqueness of rows in tables is too expensive when you do it on a per-column basis, we choose to check uniqueness based on one column only. We choose this column and make it the primary key. As soon as you make a column the table's primary key, the database will start using this column to check uniqueness. As long as this column holds unique values, no value is stored twice or more. The rows are said to be unique.

The primary key is a candidate key chosen by the database developer to be used by the database for checking the uniqueness of entire rows.

An important choice when designing databases is the selection of primary keys. This choice impacts the performance of the database. Besides performance, it also has an impact on how well uniqueness is actually enforced, not from a technical perspective, but from a business perspective.

Foreign keys

Before we go into the details of choosing a proper primary key, we need to introduce the **foreign key**. As stated earlier, a database will most likely have multiple tables. According to the definition of a database stated earlier, the data in the database is related. This means that different tables will have relationships with each other. This relationship is created by storing the primary key of one table in another table. In this other table, it is called the foreign key. It is not a key in this table, however, as it is for the other table. This foreign key references the table where the foreign key is the primary key. Because the value is unique in the table where the column is the primary key, each value in the foreign key references a specific, unique row in the table where the same column is the primary key.

To make this clear, have a look at *Figure 1.5*. Two employees are being paid monthly salaries. For the employee **Peter**, there are three payment records, one for each month for which he received a salary, while **Janneke** has been paid a salary twice:

EmployeeID (int)	Name (nvarchar(50))	Date (date)	Salary (money)
1	Peter	1-jan-2020	5000
1	Peter	1-feb-2020	5000
1	Peter	1-mrt-2020	5500
2	Janneke	1-jan-2020	6000
2	Janneke	1-feb-2020	6000

Figure 1.5 – Salary payment

Instead of storing everything in one table, we could store the same information in two separate tables as well. All the details about a person that have nothing to do with the salary they receive are stored in one table, while everything to do with actual payments is stored in a separate table. That leads to the design of *Figure 1.6*:

EmployeeID (int) (PK)	Name (nvarchar(50))
1	Peter
2	Janneke

EmployeeID (int) (FK)	Date (date)	Salary (money)
1	1-jan-2020	5000
1	1-feb-2020	5000
1	1-mrt-2020	5500
2	1-jan-2020	6000
2	1-feb-2020	6000

Figure 1.6 – Normalized salary payment

In *Figure 1.6*, the second table has a column called **EmployeeID**. This column does not store unique values because we enter a new row in this table for every month that a person receives a salary. For each month, we have to store the same value for **EmployeeID**. In the first table, however, it uniquely identifies each employee. Each value is used only once and can be a perfect fit for the primary key. In the second table, it is used as a reference to a row from the first table, making it a foreign key.

A foreign key is a column that references the primary key in another table, enabling us to combine rows from two tables. This enables us to divide our data over multiple tables without losing the relationships that exists between data elements. Splitting tables is the basis of normalization.

Choosing the primary key

The primary key plays a crucial role in databases. Logically, it is the characteristic we use to make each row unique within a table. It enables us to always select individual rows to work with. This means that we can work with unique customers, unique products, and so on. This is a hard necessity when you implement a webshop, for example. Consider a webshop that cannot distinguish one product from another, meaning they just send you a random product. Would you shop at that webshop?

As a database developer, you choose the primary key from all the candidate keys (including a surrogate key that you may add). For years, people debated whether a logical key or a surrogate key was the better choice. In Data Vault modeling, you preferably use business keys, which is a logical option. Almost all other modeling techniques prefer surrogate keys. Surrogate keys have a couple of advantages over business keys:

- They provide better stability.
- They never need to change.
- They provide better performance.

Surrogate key advantage 1 – Better stability

Imagine you had to set up a Dutch database before the year 2007 and you needed to design an `Employee` table. Prior to 2007, the Netherlands used SOFI numbers, which are equivalent to SSNs. It was obligatory to store this information regarding your employees. Since every person in the Netherlands had a unique SOFI number, this is a candidate key and you could have used it as the primary key.

Each table that has a relationship with the `Employee` table, for example, the `Salary payment` table, uses the SOFI number to keep track regarding which payment belongs to which employee. The SOFI number is now used in two tables. A primary key in general has a big chance of being used in several other tables as a foreign key.

In 2007, the Dutch government decided to switch from SOFI numbers to BSN numbers. To make this change in your database, you now need to change all columns' SOFI numbers to columns storing BSN numbers. However, this is far more complex than changing "ordinary" columns. You need to not only change the column in the `Employee` table, but also the columns acting as foreign keys in other tables. This takes time and you run the risk of something going wrong, making your data invalid or inconsistent. You do not want to end up in a situation like this, but you cannot stop the government from changing the rules.

With a surrogate key, you will never have this kind of problem. No government can force you to change your own meaningless values that you use as keys. Even new legislation such as the GDPR cannot hurt you because your key values are meaningless and, by themselves, can never provide you with any privacy-related information. If you use initials as primary keys, however, you will have problems. Initials may lead you to the actual person. There is meaning in them. Simple numerical values are the best choice.

Surrogate key advantage 2 – No need to change

Another type of stability is the fact that surrogate keys never need to change. You cannot guarantee the same for business keys. What if you use SSNs as primary keys and, after a year, you find out that you made a typo when entering a person's SSN. You will have to change the value to the correct SSN, again both in the primary key as well as in all the tables where the SSN is now used as a foreign key.

Apart from having to change the same value multiple times (which takes up precious time), the database will have to check the uniqueness of the new value. So, changing a column that should hold unique values is more expensive than changing an "ordinary" column. In Azure, *expensive* translates into both time (it will take longer, which is bad for performance, but also potentially for concurrency) and money (you may need to assign more compute power to your recourses).

Whenever a value is meaningless, it is also meaningless to change the meaningless value into another meaningless value (I love this sentence). Using surrogate keys avoids these problems.

Surrogate key advantage 3 – Better performance

The most important argument for using surrogate keys is the overall performance of the database. A database with perfect data but slow performance will certainly be a failure when used in everyday applications. A database needs to perform well, especially when it is an operational database, such as, for instance, the database of a webshop. Apart from the performance of individual queries, we also need to worry about scalability. Will the database still perform to an acceptable level when, in a year or so from now, we have a lot more data in it and/or we have many more concurrent users? Bad performance will always hurt scalability as well.

When using a surrogate key, the database developer gets to choose the data type of the column used as the primary key. Computers work fastest with numbers. Numerical values are small in memory and easily stored as binary values. Business keys are often bigger in terms of how many bytes they need, which makes surrogate keys more efficient.

This argument is important on three different levels:

- Performance when loading data or entering new data
- Performance when joining data
- The actual size of tables

Performance when loading data or entering new data

For each new row that is entered in the database, the key needs to be checked in terms of uniqueness. That check is faster (takes less time and compute) with a small efficient key. This means that the maximum possible insert rate on adding data to a table increases. Besides this, entering new rows will be faster when the key values are stored in ascending order and new rows have higher key values than already existing rows. This happens to be exactly the case with surrogate keys when you let the database generate values to be used as primary keys.

> **Note**
> Loading new data in a table is faster when using a small primary key with always increasing values.

Performance when joining data

As already mentioned, a primary key is used in other tables as a foreign key. Very often, you will have to combine tables into a single result set when using the data. You might, for instance, need to combine name and address information from an `Employee` table with salary payment information to create an annual statement. You need to combine both tables using the keys. The database will compare all the values in one table with all the values in the other table and combine those rows with matching values.

Comparing two values is not a problem for a database. However, if your tables have millions of rows, this may lead to two values being compared millions of times. And when multiple users concurrently query the database using joins, this becomes even more problematic. When your key is 128 bits in size instead of 64 bits, these comparisons take twice as many CPU cycles. That either means that the performance is going down, or it means that you have to spend more money on a service tier that provides you with more CPU capacity. With business keys, you are dependent on the size of the actual data, while with surrogate keys, you always choose integer data types that can fit within 64 bits. Multiples of 64 bits are important since we work with 64-bit computers.

If you work with relatively small databases with only a couple of users, making "bad" choices will most likely not harm you (too much). Making proper choices at design time is no more difficult or expensive than not caring. Changing an already operational database that already stores data is always expensive. It might mean that you have to take the database offline for some time. It may even mean that some code might break. And databases tend to grow. Make sure your database will scale with that growth. A lot of minor performance enhancements may have a huge impact when combined.

The actual size of tables

Some tables store millions, or even billions, of rows. Suppose a table of that size holds a foreign key referencing another table. A supermarket, for instance, might have a table with a row for each individual product scanned at the register. Each row contains information about the product being scanned, the price of the product, the date and time of the transaction, and the person operating the register. How many new rows would you get if this is a supermarket that operates nationwide?

In this example, you could store the product name of the product being sold in each row or just a product number. So, you store something like "Peanut butter," or just the number **1**, assuming that in the **Product** table, the product with the name "Peanut butter" has **1** as its product number. The name in this case is 13 characters long. That would normally take 26 bytes to store in SQL Server. If you choose int to be the data type for the **ProductNumber** column, you need only 4 bytes to store the product numbers. When you use **ProductNumber** as the key instead of using **ProductName**, you would save 22 bytes per row. With a billion rows, that equates to 22 billion bytes or 22 GB.

Inefficient keys make databases (much) larger than they need be. In Azure, extra storage immediately leads to additional costs. It also means that your database server will most likely need more memory. You could run into memory-related performance issues a lot sooner. And like I said, changing the database later is far more difficult and costly than doing it right the first time.

With Data Vault as an exception, most people tend to agree that using surrogate keys is far better than using business keys. One downside of using surrogate keys is that you add extra meaningless data to your database. This extra column makes the table a bit larger and a bit less efficient. However, as explained above, the effect on the tables, where the key acts as a foreign key, is much greater, far outweighing this argument.

A surrogate key has one real disadvantage. Each time you enter a new row, the database automatically generates a new unique value. However, that means that there is no longer any real check on whether your actual row values are still unique. All columns except the surrogate key could have the same value, which essentially means you now have duplicate rows. Technically, you have unique rows; from a business perspective, you have duplicates.

All arguments considered, using surrogate keys is better than using business keys in almost all scenarios.

Surrogate keys are a better choice than using business keys for most tables in most databases. Data Vault-modeled databases are the exception to this rule.

Integrity

Now that we know how to choose primary keys, let's look at what foreign keys are for. They provide integrity for our data. Good data is critical to businesses, so let's look at how foreign keys can help to improve data quality.

Primary keys and foreign keys make working with data in databases straightforward when the data is stored in more than one table. By using primary keys, we ensure that we can read, change, or delete individual rows from tables. You don't necessarily need primary keys to perform these kinds of operations on data, but as explained in the section regarding the uniqueness of rows, you need the guarantee of being able to reference a unique row by a characteristic because SQL operates on sets of rows that have the same characteristic. SQL is a set-based language.

Without foreign keys, you wouldn't be able to merge data that is related but stored in different tables in a meaningful way. But when you know which column acts as a foreign key, you can always combine the data from separate tables in a single result set. All we require is that the person writing SQL queries needs to know the primary and foreign keys. It is not necessary to add that information to the database's metadata, although that is certainly an option. Have a look at the following code snippet:

```sql
CREATE TABLE dbo.Invoice
(
        InvoiceNr        int not null
      , InvoiceDate      date not null
      , CustomerID       int not null
      , SalesAmount      money not null
);

ALTER TABLE dbo.Invoice
ADD CONSTRAINT PK_Invoice PRIMARY KEY (InvoiceNr);

ALTER TABLE dbo.Invoice
ADD CONSTRAINT FK_Invoice_Customer FOREIGN KEY (CustomerID)
      REFERENCES dbo.Customer (CustomerID);
```

Lines 1 to 7 create a new table called **dbo.Invoice**. This table has four columns: **InvoiceNr**, **InvoiceDate**, **CustomerID**, and **SalesAmount**. For each column, you can see the data type of the column – `int`, `date`, `int`, and `money`, respectively. The addition `not null` means that a column should have a value. It cannot be left empty when adding a new row. This part of the code would create a table without a primary key and without foreign key(s) when executed. Creating a table is perfectly OK in SQL Server; however, it is not a best practice.

Lines 9 and 10 use the `ALTER TABLE` statement to add a primary key to the **dbo. Invoice** table. This states that the **InvoiceNr** column should, from now on, be used as the primary key of this table. When there is already data in the table, the `ALTER TABLE` statement will fail if the actual values being used so far are not unique. Once the statement executes successfully, the table has a primary key, and the database will start enforcing the uniqueness of values within the **InvoiceNr** column. Uniqueness is now guaranteed.

Lines 12 to 14 of the code in the code snippet create a foreign key that references the **dbo. Customer** table using the **CustomerID** column in both tables. When executing these lines, the database will check for each value of **CustomerID** in the **dbo.Invoice** table, irrespective of whether a matching value exists in the **dbo.Customer** table. The code fails if the check fails. Failing the check would mean that you have invoices of non-existing customers.

The real value of a foreign key is **referential integrity**. Referential integrity is the guarantee that all values in the foreign key reference existing values in the primary key. Upon entering a new row in the **dbo.Invoice** table, the database checks whether the **CustomerID** value entered exists in the **dbo.Customer** table. The same check is performed when you change the **CustomerID** value of an existing invoice. Whenever a customer is deleted from the database, the database will first check whether that results in invoices without customers. Whenever a check fails, the operation is not performed by the database. An error is thrown instead.

Referential integrity

Referential integrity is the guarantee that references to other tables using foreign keys always reference existing rows.

You could change the default behavior of throwing errors. You can add extra properties to a foreign key:

- `ON DELETE CASCADE | NULL | DEFAULT`
- `ON UPDATE CASCADE | NULL | DEFAULT`

With the addition of an ON DELETE clause to a foreign key, you specify either CASCADE, NULL, or DEFAULT. Cascade means that, in the example of the code snippet, every time you delete a customer, all their invoices are automatically deleted as well. That, of course, guarantees that you do not have invoices of non-existing customers following deletion, but you also lose the invoice information. That doesn't seem a good strategy in this particular example. Be careful with using the CASCADE option.

Specifying NULL instead of CASCADE means that the **CustomerID** column in the **dbo.Invoice** table will be made empty (NULL in database terms) instead of deleting the invoice. You retain the invoice, but you lose the information regarding which customer the invoice belonged to. In our example, this option is not possible because our **CustomerID** column is defined as NOT NULL. It cannot be empty.

The DEFAULT option is only possible when the **CustomerID** column has a so-called DEFAULT constraint defined on it. This means that the column will automatically get a value when you enter a new row without specifying a value for the column explicitly. The DEFAULT option specified on the foreign key will assign the default value to the **CustomerID** column in the **dbo.Invoices** table when you delete a customer.

The same options apply to the ON UPDATE clause that you can add to a foreign key. They apply when you change the value in the **CustomerID** column in the **dbo.Customer** table.

Both options have little to no value when using surrogate keys. Changing surrogate keys is meaningless, which means you do not need the ON UPDATE clause. And we almost never delete data from databases.

The real value of foreign keys is the referential integrity they provide. Through foreign keys, you get more consistency in your data. When your data is incorrect or inconsistent, the information you derive from it becomes incorrect. And people base decisions on the information derived from the data. Good data is critical to businesses.

A point to note is that extra checks in the database always cost some extra time and processing power. You have to balance increased data quality against (a bit of) performance loss. Most of the time, quality should win over the other arguments. But if you need the highest throughput and scalability you can get, you may choose to leave out some quality checks to achieve this goal. As a database designer, you decide where the priorities lie. You always adjust your design to the intended use and you always have to prioritize different options that you need to choose from.

Recap

From all the available candidate keys in a table, you choose one to become the primary key. When using normalization or dimensional modeling techniques, the primary key is preferably a surrogate key. The database checks the uniqueness of entire rows in a table only in terms of the uniqueness of the values of the primary key.

Foreign keys provide you with a means to combine different but related tables. This is crucial in working with relational databases. Similar techniques are used when designing NoSQL databases as well but without tables. No matter which modeling technique you use, a relational database will always consist of multiple tables that are related using foreign keys.

The Check and Unique constraints

Both the primary key and the foreign key are called constraints, as you can see in the code snippet of the previous section. They limit the values that you can store in a column. For a primary key, the limitation (constraint) is that you cannot store a value that is already used in another row. For a foreign key, the constraint is that you can only use values that exist in the related table. There are two more constraints that we should mention.

The **Unique constraint** is really comparable to the primary key. A Unique constraint is applied to a column and ensures that the column can only store a value once. That allows you to use a surrogate key as a primary key and, at the same time, have the database check business keys for uniqueness.

With a **Check constraint**, you can limit the values a column can store by applying extra (but simple) rules. For instance, a **Price** column may have a money data type, making it a numerical column with precision to four decimal places. With a Check constraint, you can further specify that only positive values are allowed. Negative product prices do not make sense.

Look at the following code snippet:

```
ALTER TABLE dbo.Invoice
ADD CONSTRAINT CK_SalesAmount CHECK (SalesAmount > 0);
```

Once the code snippet here is executed, the **SalesAmount** column will no longer accept negative values.

Note

Constraints help to improve data quality in databases.

Now that we have learned about relational theory and how keys and constraints help us to improve the performance of the database and improve the quality of the data it contains, it is time to look at how we use the database. A different use case warrants a different design.

Types of workload

The performance and quality of databases not only depend on keys and constraints but also largely on the design of the database, on its schema. The chosen schema has a huge impact on the functioning of the database. In *Figure 1.7*, you can see a performance pyramid showing different aspects of a database, which is important as regards the database's performance. The basis of the pyramid is the schema. That is the main reason for this book:

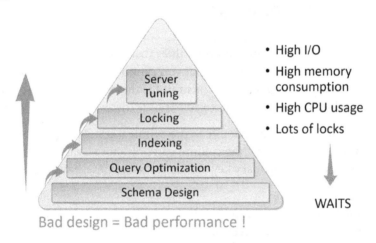

Figure 1.7 – Performance pyramid

When the schema design is good, you can further optimize a database by applying query best practices. The schema has a direct influence on how complex queries can become. Complex queries are error-prone (bad for the quality of the information derived from the data). Furthermore, complex queries are complex to execute for the database engine, resulting in poor performance.

Once your schema is correct and the queries are written according to best practices, we can further optimize the database by creating indexes. We will discuss indexes in *Chapter 4, Provisioning and Implementing an Azure SQL DB*. Writing good queries and creating indexes might be an iterative process since both have an impact on the other.

When queries are executed, you can change isolation levels. Isolation levels determine how locking and blocking work in the database engine. This balances query consistency against performance. Isolation levels are beyond the scope of this book, however.

Server tuning is a lot of things. On-premises, this might mean choosing proper hardware to run SQL Server on. In Azure, it means choosing the right service tier. Higher service tiers provide more resources (memory, storage, compute power) at a higher cost. Solving all your performance issues in this layer of the pyramid is costly. Besides, solving problems in this layer involves fixing the symptoms instead of fixing the actual underlying problem.

Issues in lower levels of the pyramid lead to issues in higher levels. Issues in the performance pyramid result in the following:

- High I/O
- High memory consumption
- High CPU usage
- Lots of locks

In the end, you either have a slow database or you are spending way too much money. Again, this is the purpose behind writing this book.

Because the schema determines what your queries look like and how they perform, the (type of) queries you need the most must be key in designing the schema. This is very true for all types of database, relational and NoSQL alike. You need to know the expected workload when designing a database.

> **Note**
>
> The expected/intended workload of a database is key when designing a database schema and choosing the proper modeling technique.

Databases are created for different types of use case. This means that different databases are used in different ways. Data Vault is a modeling technique that optimizes the database for flexible, long-term storage of historical data. You will learn more about Data Vault in *Chapter 9, Data Vault Modeling*. The focus is on storing data. Most modeling techniques focus on using the data and optimizing that usage, thereby optimizing the intended workload. There are two types of workloads (although in real life, a lot of databases are a mix of the two):

- OLTP
- OLAP

OLTP

OLTP stands for **Online Transaction Processing**. The word *online* does not refer to the internet. The term refers back to old-fashioned mainframes where you didn't interact with the system directly. You kept track of changes during the day and processed all changes during the night in batches of work. With relational databases, people started to interact with the database directly. You could "ask" (write) a query from your console and execute that query in real time.

The words *Transaction Processing* of OLTP refer to the workload. A webshop creates an OLTP workload to the database. Most **Line of Business** (**LOB**) applications do. Most applications that are part of the primary processes in a company generate OLTP workloads for their database. Besides the aforementioned webshop, you can think of **Customer Relationship Management** (**CRM**) databases, **Enterprise Resource Planning** (**ERP**) databases, and databases for financial applications or human resource applications.

Primary process databases are, most of the time, databases where new data comes into existence. New customer accounts are created, new products are entered, and (hopefully) new orders and invoices are created.

OLTP workloads have two main characteristics:

- A lot of small queries are executed.
- A lot of new rows are added to the database and existing rows need to be updated on a regular basis.

A lot of small queries are executed

Let's take the webshop as an example again. You search for a specific product. You get a list of options. You then select one of the options that looks interesting. The screen now shows a lot of descriptive properties of that product. Most of the information is about that one product. You might click on an alternative to see all the properties of that alternative. Every time you click on a product, you get all the columns of a single row. The entire `Product` table of a webshop might contain thousands of rows, but you are only interested in a single one at a time, or, at most, a couple to compare a number of alternatives.

If the webshop is a larger webshop, there will be multiple potential customers browsing through the product in the same manner. This leads to a lot of queries that query single rows, or a couple of rows at most, in other words, lots of small queries. You need the database to be fast for this type of query in order to get a responsive website.

A lot of new rows are added to the database and existing rows need to be updated on a regular basis

In the end, the webshop needs you to place orders. Maybe you need to create an account first. This means that a new row has to be entered in the Customer table. In addition, an invoice address and perhaps a different shipping address may have to be entered in the system as well, and all this to place a new order in the system that might consist of multiple order lines.

So, in addition to lots of small queries, the database needs to be able to handle lots of writes. Most of the writes insert single rows or, at most, a couple of rows at a time. Again, you need your database to be able to handle the writes quickly. Perhaps performance is less important for the writes, but scalability is an issue as well. Will the database be fast enough to handle lots of customers at the same time as your webshop grows in popularity?

Databases that focus on OLTP workloads should be normalized. You will learn how to normalize a database in *Chapter 3*, *Normalizing Data*.

OLAP

OLAP stands for **Online Analytical Processing**. Analytical processing means that we use the data for data analysis or analytics. Perhaps we are trying to answer questions such as what type of customers buy what type of products? What colors are most popular for my products? What else might a customer buy whenever they buy a product? We use this to create recommender systems. Most webshops use recommender systems nowadays. Webshops often say stuff like "People who bought this were also interested in…".

Another example might be straightforward sales reports that show how well we are doing from a sales perspective or an analysis of sick leave among employees.

All these types of queries are considered to be OLAP queries.

OLAP workloads also have two main characteristics:

- The workload is (almost) read-only.
- Most queries use large datasets.

Read-only

OLAP is about reporting and analysis. Most of the time, it is not part of the primary process. With OLAP, we don't create new data. We use the existing data that comes from primary processes. We use data to analyze. That means we read the data; we do not change the data or add data. That is to say, a process will write new data to the OLAP database and then users will use the data by only reading from the data.

Large datasets

Especially for something such as creating a recommender system, you need a lot of data. You analyze all orders from the last couple of months (or even years) to find products that are often purchased together. You are not interested in single individual orders. You are not interested in single products. You need a lot of orders to ascertain whether certain products are often purchased together. You read lots of rows from the database to do this.

A simple sales report showing sales by month and the growth in sales over time also requires lots of rows from the database. A sales report with year-to-date figures and growth compared to last year requires at least 2 years' worth of orders. That is a large dataset that you read from the database.

The report should be quick to display, especially when salespeople look at the report multiple times per day. They don't like to wait for a minute to see their report. This means that the database should be fast in retrieving these large datasets.

Databases that are OLAP-oriented often benefit from dimensional modeling. You will learn about dimensional modeling in *Chapter 7, Dimensional Modeling*.

Summary

Our current society thrives on data. People sometimes say, "Data is the new gold." Even though most of the digital data we create is not stored in databases (think about movies, music, pictures…), databases play a crucial role in our information-driven world.

It is important to consider what you want to do with your data and what the nature of your data is. Choose the right system for the job, whether that is a NoSQL database or a relational database. Choose the proper modeling technique to set up your database.

In this chapter, you learned mostly about relational databases, although much of what has been covered applies to databases in general. A relational database is a database based on set theory, where data is stored as rows in tables. Rows should be unique in tables. We use a primary key to enforce uniqueness, and we use foreign keys to relate tables to one another.

The best way to design a database depends on its intended usage. We will cover this in detail in *Chapter 3, Normalizing Data*, and in *Chapter 7, Dimensional Modeling*.

The next chapter focuses on entity analysis. Entity analysis is about what to store and how to store it.

2
Entity Analysis

In *Chapter 1*, *Introduction to Databases*, you learned why we use databases. The main concepts behind databases were introduced as well. Now it is time to take a closer look at what we store in databases and how we do so.

With big, complex systems, diving into the details from the start is a recipe for disaster. You won't be able to see the wood for the trees. This is true for database design as well. In the end, you will need to consider all the nitty-gritty details. However, start with the bigger picture first. That is the main goal of this chapter, to start from a high level and use a top-down approach to learn about the system before diving into the details of the design.

The chapter will start with (a lot of) theory and will end with an example.

We will discuss the following topics in this chapter:

- Scope
- Understanding entity relationship diagrams
- Entities
- Relationships
- Creating your first ERD
- Context of an ERD

Scope

When setting up a database, you need to know and understand the data you are going to store. You also need to decide on what data is relevant to you and what you are not going to store. You can perform an entity analysis to both define what is relevant to you and understand which parts of the data are relevant. Entity analysis also provides you with an insight as to how complex your data is, which provides you with crucial insights into the risks and timings of your project.

Entity analysis is a top-down approach that provides an initial, high-level view of what our database will look like. Lots of details about the database and the data to be stored will be missing. Along with these details, a lot of the complexity will still be missing. But performing entity analysis enables us to learn about the intended database. We will get a feel for what should and should not be a part of the database. We might even still need to get a first impression of how complex the project at hand is.

Top-down means we start by trying to get a general understanding of the process we are designing a database for. From a generic viewpoint, we deduce roughly what tables we will need to create – or, in the case of a NoSQL database, what entities we need to store information about. In relational databases, entities translate to tables, whereas in NoSQL databases, that is not necessarily true. You probably don't have tables to begin with. An important goal of entity analysis is to determine scope.

> **Note**
> Scope defines what we need to start a project and what functionality the project should deliver.

There are two types of scope:

- Project scope
- Product scope

Project scope

Project scope is the work that needs to be done to create a product or service (in our case, a database) that adheres to the demands of the business.

A database by itself is nothing. It is basically just a filing cabinet filled with dossiers holding information. If nobody opens the drawers to add new dossiers or to read the information on the existing dossiers, you may as well not have the filing cabinet at all.

To translate that back to databases, you might have to create an application that uses that database or you might need to decide on a tool such as Power BI that you will use on top of the database. In the latter case, your database might be a reporting database that you need to optimize for either Power BI or the **Extract, Transform, Load** (ETL) process that loads data into your database. You need to know what tasks need to be completed in order to create a useable database. You will read more about ETL in later chapters.

The way it is described here makes this all sound like an old-fashioned waterfall approach. But using, for instance, Scrum in an agile development environment still means you need to estimate what you need to do and how complex that is. Particularly, you need an estimation of how much time it will take. How many man-hours and/or Scrum sprints do you need to create a database or to extend a database with new functionality? When you understand the processes using your data better, you will be better able to estimate the complexity and time needed to design and implement the database and possibly the processes around the database, such as ETL processes or data migrations.

Product scope

Product scope is a description of the demands that a product (database) should meet. What should be part of the database and what should not? Once you start modeling, you can model the entire world. Of course, that is useless. You need to know what should be part of your database model and what should not. What functionality is critical and what will be easy to add later? The database is very often the fundamental aspect of whatever you are building. With the fundamentals, you often get one chance only. Whether you are using a waterfall-like method or you are more agile, you need to be able to estimate how much time you need. When you build more functionality, you need more time. Defining scope means limiting what you build, meaning limiting the time needed to build it.

Entity analysis will help in determining scope. We have to introduce some concepts and terminology first. Without these concepts and terminology, we will not be able to go through the example. The first thing we need to learn about is what an **Entity Relationship Diagram** (ERD) is.

Understanding entity relationship diagrams

By performing an entity analysis, you analyze a process or a part of a process. You focus on the data used in that process. What data do you need? Take, for instance, a planner. What do they need to know to be able to do the planning? In addition, you look at the results of the process. You store the results of whatever was planned.

As another example, a store needs to keep track of which products will be sold for what price. The sales department needs to know which customers buy what kind of products and for what amounts and what the costs are to be able to sell those products. All this information needs to be stored somewhere in the database. Looking again at an online shop, you need to store data about customers, products, and orders. These are the entities involved in this process. Entities are specific objects in the world you want to store data about. Entities will be defined in the next section. Entities become tables in our databases because they are what we need to store data about.

> **Note**
>
> An entity becomes a table in a database because data needs to be stored about entities.

The result of an entity analysis is an **ERD**. *Figure 2.1* shows an example of an ERD. It shows the Northwind database. Northwind is an old example database that was used by Microsoft a long time ago. Because of its simplicity, we will use it in this book from time to time:

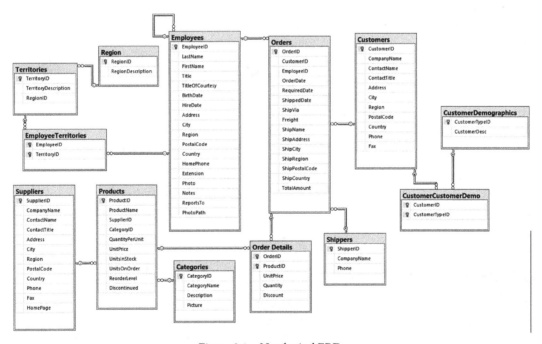

Figure 2.1 – Northwind ERD

An ERD shows the tables of a database and their relationships. This makes sure that they serve as documentation of a database as well. When you write SQL queries against a database, you need to know the tables and their relationships. Using an ERD is an effective way to get that information. That is why ERDs are used a lot.

Within an ERD, each table (meaning each entity) is pictured as a rectangle. At least the name of the table is shown in the rectangle. Sometimes, more information about the table is shown, such as what the primary key of the table is. This can be useful because you need to know the primary keys and foreign keys when querying the database. In the example of *Figure 2.1*, we even see a list of all the columns in each table. That is beyond the goal of entity analysis. Remember that we start with a top-down approach to get a first impression. For now, we'll skip the details of exactly which columns will be part of a table. For now, knowing the entities (tables) and their relationships is enough.

ERDs show tables and relationships. These relationships between tables are based on foreign keys and are visualized with a line between the tables. Foreign keys were discussed in *Chapter 1, Introduction to Databases*. In *Figure 2.1*, you can see an infinity sign (∞) at one end of the relationship. The other end has an icon of a key, denoting the side of the primary key.

> **Note**
> The goal of entity analysis is to create an ERD of a database based on a generic description of a process.

Entities

According to Merriam-Webster, an **entity** is "something that has separate and distinct existence and objective or conceptual reality." A more practical description would be something that you want to store characteristics about. When you sell products to customers, you need to keep track of certain characteristics or properties of those products. At a minimum, you need to know the price that the product sells for. So here, the product is an entity and the price is a characteristic of the entity. **Attribute** is another often-used word for a characteristic or property.

Entities translate easily into tables. For each entity that plays a role in a process, we create a separate table in the database. Each characteristic or attribute that you need to keep track of will become a column in the table.

Using the term *entity* is not completely accurate (but it is a term in everyday use). Most of the time, we mean entity type when we say entity. An entity type is a generic concept; for instance, author is an entity type. Each person who writes a book is an author, thus they are an entity of the **Author** type. An entity is the real instance of something or someone of a specific entity type. I, Peter ter Braake, am an entity of type **Author**. See *Figure 2.2* for another example:

Student				→ Entity Type

Roll_no	Student_name	Age	Mobile_no
1	Andrew	18	7089117222
2	Angel	19	8709054568	→ Entity
3	Priya	20	9864257315
4	Analisa	21	9847852156

Figure 2.2 – Entity versus entity type

A table in a database is equivalent to an entity type. In a library database, we may create an **Author** table. For each "real" author that we have books of, we will write a row (a record) in the table. Each row describes an entity.

This chapter should have been called *Entity Type Analysis*. We are not interested in entities yet; we are interested in entity types. Which real entities there are, meaning what actual rows we will store in the database, is of no consequence yet. In everyday life, we say entity when we mean entity type. We will use both terms interchangeably throughout most of this book.

You can describe an entity type in terms of what you know about the entity type. You know the name and price of products. That makes both the `Name` and `Price` attributes of the `Product` entity type. For each entity, these attributes have values. For instance, you might have a "real" product, meaning an entity where the `Name` attribute has the value `Peanut butter` and `Price` has the value `$1.50`. A row in a table stores all the attribute values of an entity in the columns of the table.

> **Note**
> An attribute of an entity is a characteristic of something or someone that exists that you need to keep track of. In a database, that translates into columns of a table.

The attributes define the entity. When two entities have the same values for all the attributes, the two entities are in fact the same entity. When products have unique names, we can distinguish one product from another. To be absolutely sure that we can distinguish each product from all other products, we might add a unique number to its characteristics. Sound familiar? Adding a price to the characteristics now provides us with enough information to start selling products. We created an entity (table) called **Product** with the **ProductID**, **Name**, and **Price** attributes (columns), as you can see in *Figure 2.3*:

 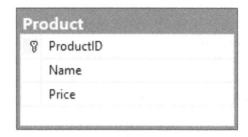

Figure 2.3 – The Product entity

Doing entity analysis, we usually only draw a rectangle with a name in it, as you see on the left-hand side in *Figure 2.3*. We need to realize at this stage that we need a **Product** entity. What attributes we need for **Product** will come later. A solid definition of the entity, however, is necessary. That might be intuitive for an entity such as **Product**, but what about a **Customer** entity? Is a customer a person or a company, or can it be either? Answering this question tells you something about the process. Getting to know the process is a big part of why you undertake entity analysis.

Knowing all the attributes is not necessary at this time. At the same time, the attributes define the entity. Companies have different characteristics other than people. So, whenever you are trying to figure out a good definition of an entity, thinking of useful attributes that will be part of the entity helps to figure out what exactly you are talking about.

So far, we have talked about the `Product` and `Customer` entities. Are both in fact entities? Of course they are, but are they of relevance to you? Or in other words: what is the scope of the project? When you are building a point-of-sale system in a regular store where anonymous people pay for products they purchase, `Customer` is not a relevant entity. When you create an online shop and you need to store address and billing information, `Customer` is a relevant entity.

For the regular store, it might be interesting to ask further questions. A lot of stores nowadays have customer cards. You subscribe to their customer card program in exchange for discounts. When you subscribe, you provide the store with personal details that they will store in their database. All of a sudden, the `Customer` entity becomes relevant for these regular stores.

Asking questions about the process and the goal of the database is an important part of why you want to do entity analysis. It forces you to ask questions. You will probably not be able to draw a correct ERD when you don't know enough about the process to be modeled. Asking questions now will enable you to design a better database later.

Understanding super- and sub-entities

Once you really dive into understanding a process, you start to see more nuances. Are a salesperson and a teacher delivering courses to customers in a company the same? They both get a salary and can both be seen as employees. The salesperson probably has a relationship with the customers, and a teacher has a relationship with the students. The salesperson sells courses to customers. In a business-to-consumer model, Customer and Student are likely to be the same. In a business-to-business scenario, however, the customer is a company whereas the student is an employee of that company.

Figure 2.4 shows an example. **Employee** is a so-called super-entity. Both **Teacher** and **Salesperson** are entities. They have some characteristics in common, such as, for instance, Name, PostalCode, and City. They both have a relationship with the **SalaryPayment** entity because they get paid salaries:

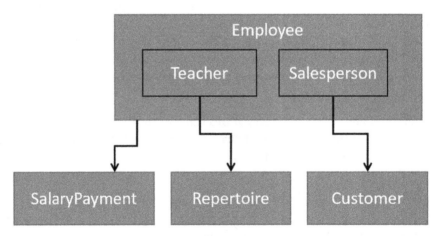

Figure 2.4 – Super- and sub-entities

At the same time, some possible characteristics are different for **Teacher** and **Salesperson**. For a teacher, you need to keep track of which courses they are capable of delivering. In *Figure 2.4*, you see that in the form of a relationship to **Repertoire**. A salesperson doesn't have a repertoire, as they are responsible for a set of customers. **Salesperson** has a relationship to the **Customer** entity.

The **Teacher** and **Salesperson** entities are so-called sub-entities.

You should recognize sub-entities whenever they are present. They are difficult to implement in relational databases but not so much in NoSQL databases. In relational databases, you cannot create a table within a table even though *Figure 2.4* suggests such a construction. Maybe you later decide on an **Employee** table with a **Function** column. Or maybe you will create separate tables for **Teacher** and **Salesperson**. Once you have more detailed knowledge of the company under investigation and the differences between teachers and salespersons in particular, you have to decide. For now, your understanding of the business model has increased by creating a high-level ERD.

Naming entities

Before we continue and focus on relationships, let's spend some time on naming entities.

For starters, clear names help tremendously in understanding an ERD. The definition of a database in *Chapter 1, Introduction to Databases*, states that a database is a self-describing set of data. A name is just a name. At the same time, calling a table X1 is not self-describing, let alone self-explanatory. Tables with names such as Customer or Invoice are a lot clearer.

Secondly, a good name means that people interpret the ERD better. Attributes define the entity, but with a clear name, the intention should be intuitive.

There has been a lot of debate about whether a name should be singular or plural. Is it Product or Products? In the end, it doesn't matter one way or the other. Being consistent in your choices is probably the best guidance.

Formally, singular is better. A table is the implementation of an entity type and each row is a single entity. In other words, the Product table is an implementation of the product concept in the real world. Each row describes a specific single product by storing the values that make it unique. Because a table represents a single concept, a name in singular is better.

As already said, a name is just a name. Use clear names for entities that make the intended definition intuitive. Be consistent in how you name your entities.

Relationships

ERDs are not solely about entities but are also about their relationships. Let's return to the example where we sell products to customers. Do **Product** and **Customer** have a relationship?

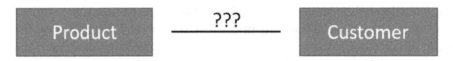

Figure 2.5 – Is there a relationship between Product and Customer?

At first glance, you might think **Product** and **Customer** do have a relationship. A customer buys your products. Nouns are often good indicators of entities, while verbs are indicators of relationships. "Buys" seems to suggest there is a relationship. However, we need to look closer. What if a customer buys a product and next week they come back to buy the same product again? We probably want to keep records of both events, both sales transactions, individually. They might fall in different calendar quarters, meaning there is a difference for you in terms of payments of VAT to the government.

There is a third entity: Sales transaction or Order. Every time you log in to an online shop to order something, a new order is made. In a regular shop, a new purchase is registered. Products have relationships with customers through the Order entity.

Types of relationships

In theory, there are three types of relationships:

- One-to-many relationships
- One-to-one relationships
- Many-to-many relationships

One-to-many relationships

One-to-many relationships exist when one row (record) from one table is related to multiple rows in the related table. Looking at it another way, each row from the many side is related to just a single row in the related table. A relationship is bi-directional and you must consider both directions. Traversing the relationship in one direction brings you to (possibly) more rows, while traversing the relationship in the other direction leads you to a single row.

This might seem abstract but a simple example can clarify a lot. Each order in an online shop is placed by a single customer. So, each order is uniquely linked to a single customer. Most online shops, however, hope to have returning customers. In that case, a customer has placed more than one order over time. Each time a customer returns and places a new order, a new row is entered in the `Order` table. This row is uniquely linked (belongs to) the row describing the customer. The previous order was linked to this same row because it was placed by the same customer. One customer has multiple orders but each order belongs to a single customer. One-to-many.

Whether customers have really placed multiple orders is not the point here. It is all about the possibility. Even when a specific shop has never had a returning customer, each customer is linked to just a single order. But will we refuse a customer when they return or will we allow the customer to place another order? The possibility of the latter alone makes this a one-to-many relationship.

You already learned that relationships are implemented using foreign keys. The primary key of one table is added as an "ordinary" column in the related table. The primary key has the characteristic that it should always store unique values. For the foreign key, no such demand exists, so it can store the same value multiple times. The same customer ID can be added to the `Order` table multiple times. That makes the foreign key the many side and the primary key the one side of the relationship.

In some literature and in some applications, you can encounter many-to-one relationships. However, they are identical to one-to-many relationships. It is just seen from a different perspective. Is your reasoning from one table to the other or vice versa?

One-to-one relationships

In **one-to-one relationships**, one row from one table relates to a maximum of one row in a related table and vice versa. You don't see this often. According to relational theory, these tables should probably be merged into a single table.

You might even say one-to-one relationships do not exist. Or, phrased in a better way, a one-to-one relationship is a special case of a one-to-many relationship. You could set a maximum on what many means. Suppose you set that maximum to the value 1. You now have a one-to-one relationship.

In *Chapter 1, Introduction to Databases*, you learned about UNIQUE constraints. Using a UNIQUE constraint on a foreign key makes sure that you can use the same value in the foreign key only once. This changes your ordinary one-to-many relationship into a one-to-one relationship. Both types of relationships are implemented using a primary key with a foreign key combination.

In the section about entities, we briefly talked about super- and sub-entities. One-to-one relationships could be used in scenarios with super- and sub-entities. Suppose you have a taxi exchange company. You keep track of the name, date of birth, and phone number of all employees. Employees can be divided up into taxi drivers and office personnel. For drivers, you need to keep track of some additional information, such as their driver's license, expiry date, and type of license. `Driver` is a sub-entity (or a specialization) of `Employee`.

One way to model this scenario is to create an `Employee` entity with an optional one-to-one relationship to the `Driver` entity. Each employee is stored as a row in the `Employee` entity (table). For each driver, we additionally store a row in the `Driver` table with driver-specific characteristics.

Another way to model this scenario is to create one big table with both columns that apply to everyone as well as columns that only apply when a row describes a driver. Now is too early to decide which solution is best. You need to know more, such as, for instance, how many of the employees are drivers, how many columns are specific to drivers, and how often you query for driver-specific information together with more generic columns. You will learn more about this in *Chapter 4, Provisioning and Implementing an Azure SQL DB*.

Many-to-many relationships

Lastly, we have the **many-to-many relationship**. Have a look at *Figure 2.6*. When you buy new shoes at an online shop, you might buy some new socks to go with the shoes. Your order comprises two products. The shoes and socks together are the order that is shipped and paid for as one. There are multiple products (many) for one order. Again, the possibility that you buy more than just one product at the same time is enough for this to be a many relationship:

Order	Customer	Product
O1	John	Shoes
O1	John	Socks
O2	Jane	Shoes
O2	Jane	Socks

Figure 2.6 – Products ordered

You are not the only customer buying those types of shoes, though. Other customers are buying the same shoes. The shoes are part of multiple different orders. One product is part of multiple orders as one order comprises multiple products. So, this is a many-to-many relationship.

As you can see in *Figure 2.6*, there are two orders, order O1 and order O2. With order O1, John bought multiple products, shoes and socks. But shoes were also bought in order O2 of Jane. So one order comprises multiple products and one product can be part of multiple orders.

Many-to-many relationships pose a problem. Suppose you decide to buy five pairs of socks. We need to store the number 5 in the database. In which table should we create a `Quantity` column that can hold the number 5?

You could choose the `Product` table. But that means you can store a single quantity for each individual product. What if another customer doesn't want to buy five pairs, but just a single pair?

Choosing the `Orders` table to have a `Quantity` column to store the number of socks you want to buy is also not an option. You buy socks *and* shoes, so you need two quantities, one to store how many pairs of socks you buy, but also one for how many shoes you buy. And you need a third `Quantity` column if you decide to buy shoelaces. And so on…

The quantity doesn't specify anything about the product or about the order. It is about the combination of the two. You buy five pairs of socks in this specific order. It is a characteristic of the relationship.

Using a relational database, you always split a many-to-many relationship into a new entity that describes the relationship. You can see this in *Figure 2.7*:

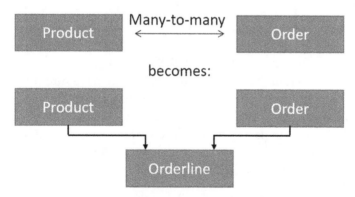

Figure 2.7 – Many-to-many relationship

As you can see in *Figure 2.7*, an extra entity (table), **Orderline**, is created. The name describes reasonably well what the entity is about. Visualize a printed version of an order. In the middle, you find a line for each product purchased. Each line has a product name, a quantity, and most likely a price.

The **Orderline** table has an `OrderID` column and a `ProductID` column. Both are a foreign key to one of the original entities that had a many-to-many relationship, **Order** and **Product**. The many-to-many relationship is transformed into a new table, often called a bridge table. The bridge table has one-to-many relationships with each original table.

> **Note**
>
> Many-to-many relationships are transformed into bridge tables that have one-to-many relationships to the original tables.

Drawing conventions

There are many ways to draw ERDs. *Figure 2.8* shows a couple of alternatives to drawing one-to-many relationships. That this list is incomplete is immediately clear when you look back at *Figure 2.1*. *Figure 2.1* was created using **SQL Server Management Studio (SSMS)**, SQL Server's management tool. SSMS denotes the one side with a key icon and the many side with the infinity sign:

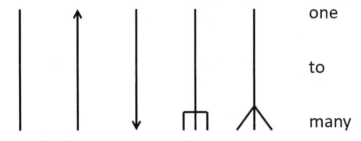

Figure 2.8 – Relationships

Figure 2.8 is only added to reinforce that you always have to make sure yourself which is the one side and which is the many side of a relationship, especially when arrows are used. In some tools, the arrowhead points to the one side. An arrow suggests it points to something (singular). Other tools might have the arrow pointing toward the many side, as in *Figure 2.7*. The drawing convention used by an application is irrelevant. You need to interpret the ERD, knowing you are dealing with one-to-many relationships.

It is important to realize that this is just a manner of drawing the ERD. An arrow may introduce confusion because an arrow might suggest direction or movement. The ERD seems to become a flowchart. But a flowchart suggests the ERD is something dynamic. Remember that a database is just a static filing cabinet. There is no directionality. Each entity stores its own data. The data is related, which is made clear by relationships between tables.

Because the way to draw relationships is different for each tool, it helps to use a drawing convention. Always drawing relationships with the one side on top and the many side underneath the one side makes ERDs easier to read. With many-to-many relationships, both entities in the relationship can be put next to each other with the bridge table underneath in the middle, as in *Figure 2.7*.

When creating an ERD, you can distinguish between optional and obligatory relationships. Optional relationships are drawn using a dotted or striped line. Technically, an optional relationship is implemented by an optional foreign key. When you allow NULL values in the foreign key, you create an optional relationship.

Earlier in this chapter, we said entity analysis is a top-down approach. You take a look at the bigger picture and try not to focus (too much) on the details. Because optional relationships are implemented with foreign keys, like every relationship, it is not really interesting (yet) to ponder too much about the nature of a relationship. Figuring out which entities have relationships and whether they are one-to-many, one-to-one, or many-to-many is important.

Recap

Logically, there are three types of relationships, 1:1 (one-to-one), 1:*n* (one-to-many), and *n*:*n* (many-to-many).

Technically, however, there exists only one type of relationship, one-to-many. This relationship is implemented using a foreign key that references the one side of the table because it "points" to the unique primary key. The foreign key is always on the many side of the relationship.

At the start of this section, we asked ourselves the question of whether a relationship exists between the **Customer** and **Product** entities. We came to the conclusion that we needed an extra **Order** entity. These entities lead us to the ERD shown in *Figure 2.9*:

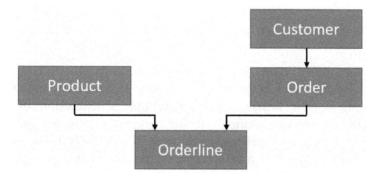

Figure 2.9 – ERD

The **Customer** entity has a one-to-many relationship with the **Order** entity. **Order** has a many-to-many relationship with the **Product** entity. This relationship leads to the introduction of a bridge table called **Orderline**. This bridge table has relationships with both the **Product** entity as well as with **Order**, linking specific products to specific orders.

Creating your first ERD

We have covered a lot of theory so far. Let's try out an example to make it clearer.

As we said, entity analysis is a top-down approach where we seek to gain insight into a process and its underlying database that stores all the data involved in the process. Take, for example, a pharmacy.

At a pharmacy, people buy medicine. Records are kept of who is using which medicines to prevent people from combining the wrong medicines. Medicines can be made from other medicines and we need to know for each medicine which components it is made from.

Customers can pay at the counter or ask for an invoice to be sent.

Let's make an ERD of this description.

One way to start is to mark all nouns and to analyze their meaning in the description. Nouns are potential entities. This is important if we are dealing with a collection (set) of possibly multiple elements and we need to know something about them. The nouns are as follows:

- Pharmacy
- People
- Medicine/Medicines
- Components
- Customers
- Invoice

Pharmacy

Let's start by taking a closer look at the noun Pharmacy. Is Pharmacy in this context a single store and are we going to set up a database for this particular store, *or* is Pharmacy a chain of stores with multiple establishments? If it is the first option, a single store, will that be the case forever or might there be more branches in the future? Is it important to keep track of which branch a transaction took place in when we have multiple branches or not?

The answers to these questions determine whether or not `Pharmacy` is an entity. When we create a `Pharmacy` entity, it will be the set of all pharmacies. Tables are sets of the same "things." If there is just one store that we call a pharmacy in everyday life, we do not have a `Pharmacy` entity in our ERD. The ERD as a whole *is* the pharmacy! If there are multiple stores, we have a `Pharmacy` entity, but only when it is relevant to keep track of store details, such as at which store a specific medicine was purchased. When there is just a single store *now* but there might be multiple stores in the (near) future, you're better off creating a `Pharmacy` entity.

To keep our example simple, we'll model a single pharmacy, meaning `Pharmacy` is not an entity.

People and Customer

In our list of potential entities, we see both `People` and `Customer`. In the context of our pharmacy, the people are the customers of the pharmacy. Because we need to keep track of who uses what, we need to store customer information. So, we do have an entity here.

The text describing a pharmacy could have been more extensive. It could have said something such as "patients come to the store with prescriptions from their doctor." `Patient` in this context is the same as `Customer`. They are synonyms for the people who use our pharmacy. It is important to recognize synonyms. You might end up with two entities (tables) that store essentially the same information. The best name for an entity depends on the jargon used in the pharmacy. Use the term the business itself uses most often.

Of course, you need to keep asking questions until you are sure that you understand the business. Are `Patient` and `Customer` really the same? What if a child uses medicine but the parent buys it? Is the child the patient and the parent the customer? Do we need to make the distinction? Maybe there are sub-entities of `People` where patients have a relationship with `Medicine` and customers might have a relationship with `Invoice`. For now, we'll continue with a `Customer` entity but we'll keep this nuance in the back of our mind.

Medicine

`Medicine` is used in singular and plural but they both mean the same thing. This refers back to the earlier discussion about naming entities.

We also encounter the word Component. In parentheses, the text itself states that a component is a medicine. Component is a synonym for Medicine in our current context. The sentence mentioning Components describes that medicines can be made from other medicines. That means that medicines have a relationship with medicines that they are made of. The Medicine entity has a relationship with the Medicine entity.

Invoice

The last noun in the list is Invoice. It could be that our pharmacist told us that their finances, including invoicing, is handled by an existing financial application they use. The new application we are building should only track which customer is using which medicines. In that case, even though Invoice is definitely an entity, it is outside the scope of our current project (database). But, depending on what the pharmacist tells you, it might just as well be an important entity for the system we build.

As we stated earlier, asking questions until you understand the process to be modeled and the scope of the project is an important reason why you should do entity analysis. By considering possible entities, you are forced to ask critical questions.

In our current example, we place Invoice inside the scope of our current database. That means we have (only) discovered three entities (so far). The next step is to investigate the (possible) relationships between the entities. See *Figure 2.10*:

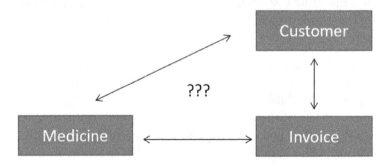

Figure 2.10 – Pharmacy – step 1

Figure 2.10 shows three possible relationships between our entities. Each relationship could be one-to-one, one-to-many, or many-to-many. Let's start by investigating the possible relationship between **Invoice** and **Customer**. Each invoice belongs to a single customer. A customer can receive multiple invoices over time. Each time you buy medicine, you are invoiced. So, after some time, you will have multiple invoices per customer in your database for most customers. We have a one-to-many relationship.

The next relationship to take a look at is the one between **Invoice** and **Medicine**. An invoice can be about multiple medicines. You can buy aspirin and sleeping pills at the same time. Both medicines are on the same invoice. Other customers can also buy the same aspirin or sleeping pills, meaning the same medicine ends up on their invoice as well. We have a many-to-many relationship. We need to create a bridge entity that we might call **InvoiceLine**, similar to the **Orderline** entity we saw in an earlier example.

With both relationships added to *Figure 2.10*, our ERD now looks as in *Figure 2.11*:

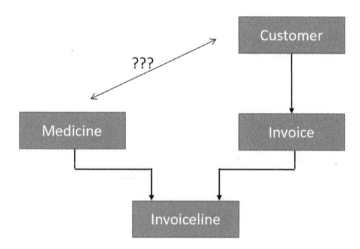

Figure 2.11 – Pharmacy – step 2

The next step is to investigate the relationship between **Customer** and **Medicine**. As you might deduce from *Figure 2.11*, they are already related. A medicine ends up on an invoice that is linked to a customer. Does that mean we do not need a direct relationship anymore?

This question is actually about how invoicing works and what we need to keep track of. We already wondered whether it matters when a child uses the medicine but the parent pays for it. Is the invoice registered to the parent or the child? But what if the invoice is registered to the parent who, at the same time as buying medicine for their child, also buys something for themselves as well? Are two separate invoices being created or does all purchased medicine end up on the same invoice? We are back at an earlier discussion: are patient and customer synonyms?

In real life, you keep asking questions until you have all the answers. In our current scenario, families can get a collective invoice. We keep track of which family member gets the invoice. The relationship that we already have between `Invoice` and `Customer` is exactly that. We need to keep track of which family member actually uses the medicine separately. We can interpret `Customer` in a broad sense. A customer does not necessarily need to get invoices or use the medicine. It is basically just the name and address details. This means we do not need to create separate entities for `Patient` and `Customer`. We could model them as sub-entities.

The second relationship between `Customer` and `Medicine` is a many-to-many relationship. Somebody can use multiple medicines and the same medicine can be used by multiple people. Again, we need to create a bridge entity. We could call this, for instance, `Prescription` because it could store which medicine a doctor prescribes to a patient and in what doses.

The ERD now looks as in *Figure 2.12*:

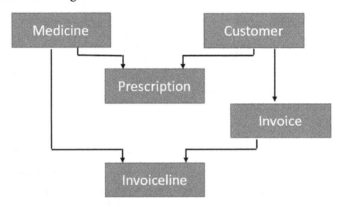

Figure 2.12 – Pharmacy – step 3

To end this example, we need to look at the relationship of `Medicine` with itself. Let's look at the word `Component`. A medicine is made up of multiple components. A component can be a component of multiple medicines. *Figure 2.13* should clarify this:

Medicine	Component	Quantity
Med A	Med X	2 ml
Med A	Med Y	3 mg
Med B	Med X	3 ml
Med B	Med Y	4 ml

Figure 2.13 – Medicines

As you can see, medicine **Med A** is made of 2 ml of **Med X** and 3 mg of **Med Y**. So a single medicine can be made up of multiple other medicines. However, medicine **Med X** is both part of **Med A** as well as **Med B**. So a single component can be a component of multiple medicines.

So, we are dealing with a many-to-many relationship. *Figure 2.14* shows how to model that in an ERD:

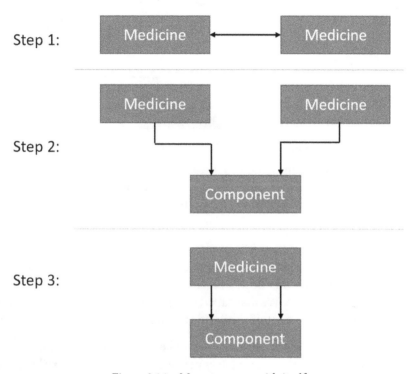

Figure 2.14 – Many-to-many with itself

In *Step 1* of *Figure 2.14*, we draw the entity twice with a many-to-many relationship. In *Step 2*, we add the needed bridge entity. In *Step 3*, we recognize that the two entities we started with are actually the same, so we draw it just once, keeping both relationships.

Adding the result of *Step 3* to the ERD of *Figure 2.12* leads to *Figure 2.15*, the ERD of our pharmacy:

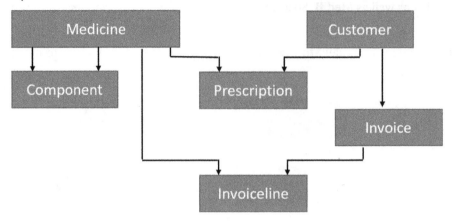

Figure 2.15 – ERD of our pharmacy

As we said already a couple of times, the goal is to ask critical questions and determine the scope to be better able to create a good, functioning database later. There are heaps of questions left unasked about the pharmacy, for example, whether it matters that people may pay in cash or ask for an invoice.

This matter is easily solved using the ERD we already have. You could add an `IsInvoiced` or `IsPaidInCash` column that indicates whether an invoice was sent or a cash payment was made. You don't need a real invoice in the real world to still use an `Invoice` entity in your database. We just called it `Invoice` because the information we intend to store in that entity has a lot of similarities with information on an actual invoice.

Another unanswered question is whether we need to keep track of whether an invoice has already been paid. If answered *yes*, a completely new world of possible complexity opens up. Do we send reminders? If so, how many? Can people pay in installments? What if people buy medicine when a debt is still open? For now, all of this is outside our scope. Just be aware that at this point, asking questions is your job.

Now that you have learned about what an ERD is and you have created one yourself, let's look into how and where ERDs are used.

Context of an ERD

ERDs are used in many different ways. We've used them here as an aid in database modeling. They can be a first, high-level step to get an impression of what we are dealing with. We define the scope and create a rough first draft of what our database might look like.

Application developers who write queries against databases need to know the database. What information is stored where? An ERD is a handy tool to get the needed insight. You see all the tables and which tables are related. When you add keys to the ERD (name an entity as we did plus add the primary and foreign keys in the entity's rectangle), you also know how tables are related. You can even add all the columns, as in *Figure 2.1* at the start of this chapter. This ERD contains all the detailed information needed to start implementing the database.

There are multiple steps to be taken in between what we have created so far and the ERD of *Figure 2.1*. In *Chapter 3*, *Normalizing Data*, we are going to add the details, creating what we call a logical or conceptual model. It uses real information needs (screens and reports and lists people work with) with all the details of a real-life example, instead of a generic description. We need to know exactly which columns are part of the tables we need to create later. We will create an ERD with more detail to it than what we have created so far.

After creating the conceptual model, we will transform that conceptual model into a physical model. We are going to implement a database as an Azure service. We need to provision the service with enough storage and compute power. So, we need to understand how big our tables will become and how many times per second we will query them. We will add even more detail to our data model.

There are lots of tools for creating an ERD on the market, from free online tools to expensive, professional tools. Most make it easy to design the ERD. Creating ERDs is an iterative process. You will never get it right first time. So, you need to be able to change and add to the ERD continuously. In the end, most tools will be able to generate the needed SQL statements to actually create the database you designed.

In short, we now have the first draft of a conceptual database design. We will work toward more sophisticated, more detailed designs in the following chapters.

So far, we have suggested that there is a one-to-one relationship between entities and tables. In NoSQL databases, you might not have tables. You do, however, need to know which entities you are dealing with and how they relate to each other. So, creating an ERD is still useful. Transforming it into a physical model will be a lot different than when using a relational database.

Summary

An entity is something that you need to store information about. Entities describe the objects, people, and processes of our everyday life. Entities are related. An ERD shows the entities of interest to our database and their relationships. In that way, an ERD reflects how we organized our processes. It gives us insight into those processes *and*, by doing so, also into our database.

In this chapter, you learned how to create an ERD from a generic description of a process.

The next chapter will add more detail to the initial database design you created in drawing the ERD.

Exercises

If you want some more practice with creating ERDs, you can use the following two exercises. You can find the solutions to the exercises in the downloads for this book.

Exercise 1 – student registration

A well-known course provider located centrally in the country needs a system to plan their courses and keep track of which students follow which courses when. They have many courses you can follow. Each course is part of a curriculum that makes courses easier to find on the website. A course may be taught by a teacher employed by the institute or by an independent contractor. For each course leader, we must know which courses they are able to deliver. We also need to be able to quickly show which courses were delivered most recently and which course a teacher delivers the most. For freelancers, they keep track of extra information, such as registration with the chamber of commerce.

Courses are not provided to private individuals. Students are registered by their employers. The employer receives the invoice. For each company, a contact person is kept. Invoicing is done by a separate invoicing application.

Draw the ERD.

Exercise 2 – airline

An airline has asked you to design a planning system for transporting individuals. The business manager describes their business as follows.

The airline has 100 airplanes of different types. A pilot is licensed to fly one type only. Flight attendants can be licensed for three different types at the same time. The airline flies to destinations all over the world. Not all types of aircraft are used everywhere. For some routes, smaller aircraft don't have the needed range. At some airports, there are restrictions around which types of planes they allow and don't allow, for instance, because of environmental restrictions.

The marketing department determines the destinations they fly to and how many passengers are expected to go to a destination. Planning needs to find an able crew and a plane for each flight. An airplane is not always available due to scheduled or ad hoc maintenance to the planes. Employees might also have other activities to complete besides flying, such as sitting exams or undergoing medical tests.

Draw the ERD.

3
Normalizing Data

Now that we are familiar with the top-down approach from the previous chapter, we will switch to a bottom-up approach: normalizing data. The goal is again to create an **Entity Relationship Diagram** (**ERD**). The difference is that we will now worry about all the details. Having learned a lot about the system to be created using a top-down approach and asking any questions as required, we will now make a conceptual model.

In this chapter, you will learn about the following topics:

- When to use a normalization strategy
- Why avoiding redundancy is important
- The formal steps of normalization
- An alternative approach to normalizing data
- Integrating separate results

When to use normalization as a design strategy

In real life, entity analysis and normalizing data may not be as separate as this book suggests. They both lead to an ERD. Hence, you can utilize entity analysis skills (arguments) during normalization, and vice versa. Especially when you don't create a new database but you alter (extend) an existing one that isn't well normalized, you will use both techniques simultaneously.

For many people, normalizing data is equivalent to relational databases. The misconception is that a database has to be normalized in order to be a relational database. In *Chapter 1, Introduction to Databases*, you learned that storing data as tables is what makes a database relational. You also learned that those tables have to be created before you can store actual data. All databases created using a **Relational Database Management System (RDBMS)** are relational databases. However, normalizing is a technique used to design relational databases.

This means that this chapter mainly relates to relational databases and is less (not) important in terms of designing NoSQL databases. In Azure, there are multiple relational databases to choose from. You can create Azure SQL databases, MySQL databases, PostgreSQL databases, and MariaDB databases as **Platform-as-a-Service (PaaS)** services in Azure. When utilizing **Infrastructure as a Service (IaaS)**, the list is unlimited.

The misconception that databases need to be normalized in order to be relational stems from the fact that, for a long time, normalizing was the only technique used to design databases. People didn't make the distinction between OLTP and OLAP, as we did in *Chapter 1, Introduction to Databases*. When Codd invented the relational database and the normalization of data, people were only building operational databases. These databases were all OLTP databases.

> **Note**
> Normalizing data is the process of designing a database schema optimized for an OLTP workload.

This means that when you want to design a relational database that will be used by an OLTP workload, it is a best practice to normalize your data. You do that after an entity analysis to get all the details you require to actually implement the database.

Considering all the details

As we said at the end of *Chapter 2*, *Entity Analysis*, a good ERD tool is able to create the entire database. This means that you need to know each column of each table (each attribute of each entity). All details must be known at the end of the normalization process.

> **Real-life example**
>
> I once got a call from a (potential) customer who needed a report showing address details of customers to be sorted on family name. That should be a matter of a simple checkmark in all the reporting tools I have worked with. Following a number of additional questions on my part, it turned out that the database had a single column for **Name**. The first name, middle name, and last name fields were stored as a single value in the **Name** column. I needed the last name to be a separate value in order to make sorting easy.

The relevance of this example is that something straightforward and common to ask for became almost impossible because of an "incorrect" database design. Which part of the name is the family name? You could try searching for a space, hoping that the part after the space is the family name. However, some people have double-barreled names, or double first names, or spaces in middle names. Just using the part after the last space doesn't give you the result you need.

Had the database been designed with a column for the first name, a separate column for the middle name, and a third column for the last name, it would have been easy to sort the report in the manner requested. By way of an additional argument, it is easy to combine two columns into a new column within a SQL query. It is a lot more complex, and it takes more time and compute power, to split a column into separate columns.

> **Best practice**
>
> Always create separate columns to compose parts instead of storing a combined value as a single value.

Designing a database is no more complex when you store a name as separate parts instead of as a whole. However, changing the database at a later stage can be really difficult. When you decide to change an existing database that already stores real data, you need to split all the values already stored. If that were easy, I would have done it on the report to begin with!

An even bigger problem is that changing the database schema might impact existing applications. You cannot guarantee upfront that applications will keep working when you apply changes to the database schema. Applications querying the **Name** column will generate errors if there is no longer such a column. You need to do an impact analysis of the proposed change. The analysis itself costs time (and money). Implementing the changes that you deem to be necessary will take even more time.

> **Note**
> It is important to get the design right the first time!

The rest of this chapter will focus on the normalizing process without focusing too much on the details described here. There are numerous examples of choices to be made. Is a name complete without a title of courtesy? Do we need an extra column for values such as "Professor"? Is the address line a single column, or are the street name and house number separate values? Is the house number a single value, or might there be additions, such as "1B"? There are numerous decisions to make. Every choice you make has an impact on what will be easy and what will be nearly impossible later on. It also has a direct impact on performance and the cost of the service tier in Azure that you need in order to implement the database.

Preventing redundancy

Let's briefly recap what the characteristics of an OLTP workload are:

- A lot of small queries are being executed.
- A lot of writes to the database are performed.

In the case of an OLTP workload, making writes (updates and especially inserts) to the database as efficiently as possible is key.

The most important premise of normalizing data is to prevent redundancy in the database. Redundancy is storing the same piece of information twice or more. We want to store each value just once as much as possible. There are three reasons for doing so:

- Redundancy costs extra storage.
- Redundancy has a negative impact on performance.
- Redundancy has a negative impact on data quality.

Let me now elaborate on these reasons in more detail.

Available storage

The first argument may seem strange in the era of big data. This argument has its origins in the past, where storage was limited and really expensive. This has become far less of an issue over recent years. However, with the cloud, you pay for what you use. Bigger databases cost more money.

The fact that in former times, storage was limited forced developers to use storage as efficiently as possible. In the 1980s, storage cost more than USD 100,000 dollars per **gigabyte** (**GB**). Nowadays, it is less than 10 cents per GB. The other side of the coin is that databases of 100 GB were considered large at the time, whereas today we create databases of tens of **terabytes** (**TB**) or even petabytes in size. A petabyte is a thousand TB, which is a million GB. With volumes like this, storage becomes expensive again.

You should always consider the costs associated with storage. Today, cost will most likely not be the limiting factor in your design decisions. However, spending money without actually needing to do so is pointless.

Performance

An important aspect of OLTP workloads is to insert new data (new rows) into the database. We all know stories about bureaucratic processes where you need to fill out forms in triplicate. These are processes where there is a lot of redundancy: you write the same information down three times.

Consider a web shop where each new order is stored in three different places. Storing a new order is now three times as much work as when you store it just once. A single order takes a third of the time without redundancy. Alternatively, you can enter three times as many orders in the database at the same time. This means that you can either have a bigger workload or you can choose a smaller service tier of the SQL database that you provision in Azure. Hence, you make it cheaper to run your workload in Azure and, at the same time, your database will scale better to larger workloads when your web shop increases in popularity.

Alongside adding new rows, you may also need to update existing rows regularly. However, the same principle holds true. Suppose you need to change a customer's address and you stored this address in three different places in your database. You need to find all three entries and change all three to complete the update. Without redundancy, one action would have sufficed.

Reading from a database is an entirely different story. Reading data from a database may benefit from redundancy, for instance, because you might have to execute fewer joins. We will see that you remove redundancy by splitting tables, which leads to more tables in the database. Fewer tables in a database means, on average, more columns per table (when you store the same number of columns). With more columns in a table, the chance that one or just a few tables hold all the columns you need increases. That decreases the number of tables you need to join.

However, redundancy in your data will probably lead to more memory usage. This is caused by the fact that the database will try to keep as much of the data in memory as possible for faster reads. That means that the data is not only stored redundantly in the database on disk, but also in memory. This could lead to memory bottlenecks and performance issues. You may add more memory to the database server to overcome this bottleneck. This means that you need to provision a higher SQL database service tier.

Data quality

The final reason for avoiding redundancy is to minimize the risk of inconsistencies in your data. Let's suppose, once again, that addresses are stored in three different locations in the database and a customer moves. You now risk changing the address in two of the locations while forgetting about the third address. You now have two different addresses for the same customer in your database. When you query the database for the customer's address, it now depends on how you write your query as to whether you get the current address or the old address. And different people will write different queries. Your database can no longer be trusted. You will read more about this in the section regarding normalization steps.

You may think that the problems may not be that bad. Unfortunately, Murphy's law applies to databases. Everything that can go wrong will go wrong. Redundancy in databases, combined with applications where people working in primary processes enter data in the database, will always lead to data quality issues in the database.

> **Note**
>
> By avoiding redundancy in a database, an OLTP workload will be faster and cheaper, while, at the same time, the data quality will increase.

How to avoid redundancy

After learning why redundancy should be avoided for operational databases, let's now look at how to avoid redundancy. Normalizing data is a process in which we split data from a report (or a form, or a screen of an application) step by step to divide the data elements between different tables. The result of the process of normalization is that we eliminate redundancy.

In our everyday processes, we work with information on a constant basis. For example, in a stockroom, people operating forklift trucks use picklists. A teacher of a course might use a presentation list. A call center employee might use a list of people to call. A support engineer might use a list of the most common issues. And when an issue is solved, they need to enter the details pertaining to the problem, the solution, and who the problem belonged to. What all these lists and forms have in common is that they show information coming from a database. It doesn't matter whether you print a list, use an app, or fill out a paper form. The information comes from, or needs to be stored in, a database.

In order to design a database, we need to know what data people work with and how they use this data. We need to know the lists and screens and forms that people will work with.

Back to normalizing data and redundancy. Data is divided between multiple tables. This leads to "narrow" tables. By "narrow," I mean that a table consists of a relatively small number of columns. If you need to store 100 different pieces of data and you do that in a single table, it will consist of 100 columns. If this data is divided between 10 different tables, you have an average of just 10 columns per table.

One advantage of "narrow" tables like this is that adding a new row to a table means adding only a couple of columns or, more technically, just a few bytes. That makes adding new data more efficient.

> **Note**
>
> When normalizing data, we divide columns between multiple tables, resulting in "narrow" tables without redundant data in the tables.

The normalization steps

The process of normalizing data is defined by a number of steps that you need to perform in order to get from a report or screenshot to a normalized table structure. Let's examine these steps in detail, starting with what is called *Step zero*.

Step zero

Data is normalized using information requirements such as reports and application screens in a number of formal steps. We will explain these steps using the report as shown in *Figure 3.1*. This report shows the hours worked by employees on different projects as of January 1, 2021. There are two active projects, P1 and P2. Two people, Peter and Janneke, are working on project P1. Peter, Jari, and Mats are working on project P2:

Project Overview

Situation on 1-January-2021
Printed by Peter ter Braake

Project	Budget	Employee	Department	Supervisor	Hours
P1 – DWH	1000	Peter	Staff	-	50
		Janneke	HR		10
Total DWH					60
P2 – Tuning CRM	40	Peter	Staff		16
		Jari	ICT	Lucy	16
		Mats	ICT	Lucy	8
Total Tuning CRM					40

Available employees:

Name:	Skills:
Peter	SQL Server, Data modelling
Janneke	Scrum master
Jari	T-SQL, C#
Mats	T-SQL, CosmosDB

Figure 3.1 – Project overview

The first step is to define a list of columns to be stored in the database. You have to consider three main points in this preliminary step:

- Can the report be broken down into separate independent reports?
- Are there parameters on the report?
- Does the report show process data?

Can the report be broken down into separate independent reports?

In some cases, we combine separate types of information on one report or screen. Because they are actually separate, you need to analyze them independently. That means you need to split the report you are analyzing into individual parts. Perform the steps as described in this section for each part independently.

Imagine, for instance, a report with the planning of a course institute. The report is a list of planned courses with a date, location, and teacher. The second part of the report contains a list of all the teachers, including the courses they can deliver. That is convenient for planning when the latter department quickly needs to find a replacement if a teacher calls in sick. The list showing who is capable of what is, however, completely independent of the list showing the actual planning. They are separate reports.

In the project overview of *Figure 3.1*, you see the actual project overview and a list of employees and their skills. These are employees that can potentially be planned to work on the projects. The second part is basically a separate report and should therefore be analyzed separately. We will continue with the actual project overview.

Are there parameters in the report?

Parameters are values in reports that do not come from the database. You need to recognize them in this step and eliminate them as columns in the database.

The project overview states that Peter ter Braake is the person who printed this report. He was probably logged in when the report got printed. The report got this from the OS and not from the database. It may be that Peter can only see projects that he works on himself. But even in that case, the name on top didn't come from the database. At this point, we are only interested in data coming from the report. Who printed the report is irrelevant, who worked on the different projects and for how many hours is the actual information that this report is about and that information needs to be stored in the database.

The name of whoever printed the report is a parameter. From now on, we will ignore this piece of data.

Above the name of the user who printed the report, a date is shown. This states that the remainder of the values on the report pertain to this particular date. In particular, the hours will probably have different values when you look at these two projects again in a month's time. Is this date a parameter even when it comes from the OS at the time of printing, or is it not?

If the report always shows the actual hours for all projects at the time of usage, the date is probably coming from the system clock and not from the database. However, it could be the case that the user using the report types in a date that may be in the past and the report shows the values from that specific date, no matter how far back in the past we look. All values should then be retrieved from the database based on this date. This is only possible when we store all values historically, meaning including a date when they were the actual values.

In the last case, you need to have a **Date** column. In the first case, **Date** is a parameter. To make our example less complex, we assume here that the report only shows the actual hours when the report is used. That makes the date a parameter, meaning we can ignore the date going forward.

Does the report show process data?

Finally, you need to recognize so-called process data. Process data is values that can be computed using other data. For instance, when you know the unit price of a product *and* you know the quantity a customer purchased of that product, the actual sales amount is just the multiplication of the two. Storing the sales amount would lead to redundancy. You could retrieve the sales amount in two different ways from the database: by multiplying the unit price by the quantity, or by reading the sales amount directly.

But what happens if someone changes the value in the **Quantity** column at a later point in time? There is a risk that the sales amount is not changed accordingly. Both ways to retrieve the sales amount will now come back with different values. Storing process data means storing more data than necessary, having more work to do when data changes, and introducing the risk of inconsistencies in your data. You need to eliminate process data from the list of possible columns.

Looking at the example of *Figure 3.1*, you can see that the database keeps track of the hours per employee. We also have the number of hours for each project. Adding the hours per employee leads to the hours per project. That makes the project hours process data.

After eliminating parameters and process data, we are left with the columns that the report got from the database. In our project overview (see *Figure 3.1*), those columns are as follows:

- **ProjectNumber**
- **ProjectName**
- **Budget**
- **Employee**
- **Department**
- **Supervisor**
- **Hours**

Notice that in this list, the project number and project name are already split into two separate columns. In the report, they are shown as one value. The project number might very well be a surrogate key in the database. We assume here that the project number is an automatically created number, a surrogate key. Concatenating the letter "P" with the **ProjectNumber** column, " - ", and the project name will make the value as shown on the report.

Notice as well that we use **ProjectNumber** and **ProjectName** as column names. Using spaces in column or table names goes against database best practices. You can use names that include spaces if you want to or need to, but we normally try to avoid it. Spaces in names increase the likelihood of errors in T-SQL code. An often-used naming convention is to start each word with a capital letter, as we did in the example.

Tip
Avoid using spaces in table names and column names.

The final part of this preliminary step we are undertaking is to regard the list of columns as a single table that needs a primary key. Basically, we want to know what the report is about. Because this report is about projects, the **ProjectNumber** column is the most suitable column for our first attempt at a primary key. We have already sort of concluded in any case that it is a surrogate key and describes the report.

Altogether, we end up with the table in *Figure 3.2* as the outcome of our preliminary step:

Project
ProjectNumber
ProjectName
Budget
Employee
Department
Supervisor
Hours

ProjectNumber	Projectname	Budget	Employee	Department	Supervisor	Hours
P1	DWH	1000	Peter	Staff	-	40
P1	DWH	1000	Janneke	HR	-	10
P2	Tuning CRM	40	Peter	Staff		16
...	...	...	...	...	...	...

Figure 3.2 – 0th normal form

Notice that we have already named the table. This is not necessary at this stage, but it might be helpful in understanding what we are doing. Also notice that we marked the primary key by underlining the column that will serve as the primary key. This is a convention that we will use throughout this book.

Notice also that the table as it stands at this stage has a couple of issues that need to be resolved. Primary key values need to be unique and, for each row, you can store one value per column. In the table as it is now, this means that each distinct project number can have only one value for the **Employee** column, or you need multiple rows with identical primary key values as shown in *Figure 3.2*. The report clearly shows multiple employees per project. That is why we need to proceed to the next formal step, also known as the first normal form.

First normal form

The first normal form looks for repeating sections in the report. Are there parts of the report that occur repeatedly? Multiple similar lines on an invoice, for example? Each line on an invoice is structured in a similar manner, with a product name, unit price, and quantity. Each line holds the same sort of information; it is just the values that are different. An invoice line is repeated for each product sold. That is what is meant by a part that occurs repeatedly.

With normalizing data (and with relational databases in general), there are only two numbers to consider: one and many. When something possibly occurs more often than once, it is many. It occurs repeatedly.

You should, at all costs, avoid creating tables with repeating columns, such as **Employee1**, **Employee2**, and **Employee3**, even if each project is always carried out by exactly three employees. What if, at some point in the future, we start a project with four employees working on it? Or five, or six, or seven…? Repeating columns do not provide any flexibility. There is always a hard limit to the number of columns you include. In addition to that, repeating columns are inefficient because often, not all columns are actually used to store values.

The first normal form takes repetitive sections and creates a separate table for a repeating section. This means that you will get extra rows instead of repetitive columns. Creating extra rows in a database is virtually limitless.

On the **Project Overview** report in *Figure 3.1*, we can see multiple projects. We can already store multiple projects using the table from our preliminary step. As long as each project gets assigned a unique project number, we can create a new unique row in our table for each new project we start.

For each project, the report shows as many lines as there are employees working on the project. Project P1 has two lines, describing a single employee each, and project P2 has three lines. The columns that (potentially) have a different value for each line are **Employee**, **Department**, **Supervisor**, and **Hours**.

The first normal form states that we put all the columns in this repetitive section in a new table. Each normal form splits tables into two new tables, which will have a relationship with one another. The four columns we mentioned are put in a new table. This table will also have the column that is the primary key of the original table we are splitting. All the columns from the original table that we did not mention stay behind in the original table, including the primary key. This brings us to the first normal form, as shown in *Figure 3.3*:

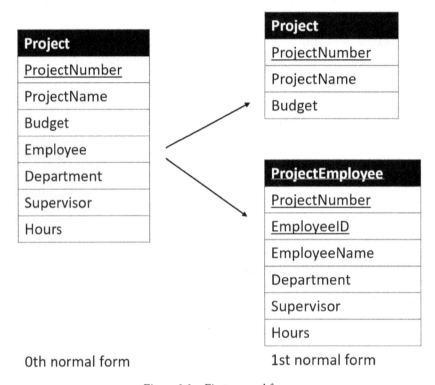

Figure 3.3 – First normal form

A column called **EmployeeID** was added to the table and made part of the key. Each line within a project is different because it is about a different employee. We can identify a line by the employee name and discriminate between two lines by the employee. That would make the **Employee** column, which holds a name, a key. We learned already that using names as keys is not a good idea. So at this point, we introduce an **EmployeeID** column and rename the **Employee** column to **EmployeeName** to make the column name more descriptive.

Combined primary key

Choosing the primary key for each table is an important step. Because the value of the primary key uniquely identifies a row, you could say that it defines the table. That is why the name of the primary key and the name of the table are often similar. One can be inferred from the other. The table with the **ProjectNumber** primary key is logically the **Project** table.

Notice that in *Figure 3.3*, the **ProjectEmployee** table has two columns underlined. Two underlined columns don't mean the table has two primary keys. It means that the values of the two columns combined uniquely identify the rest of the row. The primary key is the combination of both columns.

EmployeeID was added to the key because each line on the original report concerns a different employee. That is a different way of saying that a line can be identified by the employee that is described by that line. At the same time, there are two lines about the employee Peter. But they were printed in different sections, or under different projects. To identify a line in the report uniquely, you need to know in which project section the line is printed *and* which employee it describes. **ProjectNumber** and **EmployeeID** together uniquely identify the remainder of the values. You can see an example of the table in *Figure 3.4*.

A combined key means that the individual columns do not have to have unique values. The combined values of the columns should be unique. That is exactly what we need. We can use one **ProjectNumber** column multiple times as long as we combine it with a different **EmployeeID** column each time. Because we have multiple employees per project, this is exactly what we need.

It is true the other way around as well. We can now use the same **EmployeeID** column multiple times as long as we combine it with a different **ProjectNumber** column each time. An employee can work on multiple projects.

Each combination of **EmployeeID** and **ProjectNumber** must be unique. For each combination, you can store how many hours that employee worked on that project.

Foreign key

A second important aspect of the step described previously is that both tables have a **ProjectNumber** column. In the original table, **Project**, it is the primary key. In the table called **ProjectEmployee**, the **ProjectNumber** column is not the primary key. Because it is the primary key of the **Project** table, it is a foreign key in the other table. Both tables can be merged back into a single table, combining rows with the same value for the **ProjectNumber** column. In other words, we can join the tables using SQL.

Surrogate key

Notice that we introduced the **EmployeeID** column when we split the original table (see *Figure 3.3*). **Employee** becomes part of the key when splitting the table. Names are never unique, which makes them very bad candidates to be used as keys or as part of a key. Look at the **Project Overview** report. We see the name "Peter" on two occasions. But "Peter" is a very common name. There is no way of knowing whether the same "Peter" is working on both projects or whether both projects have *a* "Peter" working on the project but they are two different persons. Introducing an **EmployeeID** surrogate key allows us to create two unique IDs when there are two persons sharing the same name or just using a single ID for the one "Peter." Keys always have unique values.

Besides that, names are long in terms of the number of bytes, which makes them inefficient. We discussed this in *Chapter 1, Introduction to Databases*.

As soon as we start using a column as a key or as a part of a key, we need to look a bit closer at this column. Is the column guaranteed to hold unique values? Does the column have an efficient data type? When one of the questions is answered with *No*, introduce a surrogate key.

Details

On the report, the **Employee** column holds first names. It might be wise not to use the report header literally as a column name. **Employee** is too vague. We changed it to **EmployeeName** in *Figure 3.3*. **EmployeeName** is not exactly specific either. **EmployeeFirstName** might have been a better name. It explains better what to expect in this column.

Extra repeating sections

In this example, we are done with the first normal form. However, in more complex reports, there might be repeating data columns in the new table we just created. In that case, we split this table again into two tables. The new table we create in this way has a primary key that is made up of three columns. We already had two columns in the key from the first split. You add the identifying column of the repeating data to that key. All the columns that are part of the repeating group move to this new table. All other columns stay behind in the original tables, the one with the key being a combination of the two columns, which was the result of the first split.

Every time we split an extra repeating group into a new table, this table's key is expanded to include an extra column. Sometimes, adding elements to a key means that we add "freedom" to the model that does not exist in the world we are trying to model. Let's suppose that, by **ProjectEmployee**, we mean an external consultant, and let's suppose further that we never ever hire the same consultant twice. In that case, you just note for which project you hired someone. **EmployeeID** is, by itself, the key, and one and only one value for **ProjectNumber** should be stored for each value in **EmployeeID**. The **ProjectNumber** column is still a part of the table, but it should not be part of the key.

Suppose in this case that we keep **EmployeeID** and **ProjectNumber** as the combined key. You could now use the same value for **EmployeeID** multiple times, as long as you combine it with a different value for **ProjectNumber** each time. Each unique combination makes a valid key. However, this means that we can have employees working on multiple projects according to the database. But we just said that this was not possible in this scenario.

Whenever there is at most one possible value for a column for each value of the key (there is one name for each **EmployeeID**), that column is an "ordinary" column. It should not be part of the key.

Let's now go back to the **Project Overview** report. The **ProjectEmployee** table has no more repeating columns. The tables under *Figure 3.3* are said to be in the first normal form.

Second normal form

The first normal form is a big step toward a conceptual database model, but the tables may still have some issues. Look, for instance, at *Figure 3.4*. The table shows two unique rows because there are two unique primary key values, (1, 1) and (2, 1), for the **ProjectNumber** and **EmployeeID** columns, respectively. Unique combinations mean unique values for the key, which means valid rows as far as the database is concerned. But what is the name of Employee 1? Is it "Peter" or is it "Paul"?

Project-number	EmployeeID	EmployeeName	Department	Supervisor	Hours
1	1	Peter	Staf	-	50
2	1	Janneke	Staf	-	50

Figure 3.4 – Possible issues after the first normal form

The problem with the table indicated in *Figure 3.4* is that it still holds redundancy. **EmployeeID** was introduced to uniquely identify an employee. **EmployeeID** has nothing to do with projects. An **EmployeeID** value of 1 does not mean the first person to work on this project; it is just the unique number identifying an employee. Suppose both rows in *Figure 3.4* had shown the value "Peter." The name "Peter" would then have been stored twice. Storing the name twice is inefficient *and* means that we run the risk of making an error, such as having the name "Paul" in the second row even though the **EmployeeID** value is the same, which means it should be the same person. There is now an inconsistency in the data. Asking the name of employee 1 seems a logical question. However, you cannot get an unambiguous answer.

The issue described here arises from the combined key. The combined key means that **EmployeeID** is no longer unique in this table, even though it should uniquely identify an employee. This is the type of issue that the second normal form resolves. The second normal form looks for complete functional dependencies.

A complete functional dependency means that whenever the key value changes, all values of all other columns *could* change as well. Values in ordinary columns do not have to change because they do not need to hold unique values. But if the value of a column *may not* change when the key *does* change, it is placed in the wrong table.

In the example of *Figure 3.4*, the name cannot change when, in the key, only the value for the **ProjectNumber** column changes. The same person working on a different project is still the same person with the same name. The name does not depend on the project number. The name only depends on **EmployeeID**.

Two values are said to be completely functionally dependent when one value follows from the other value. Let's take the date "April, 2021" as an example. It is now unambiguously clear that the year we are talking about is 2021. The year is functionally dependent on the month. The month could be the key for the year. Notice that this is not true when you define the month as just "April." From "April" alone, you cannot deduce the year because there is an April in 2020, but there is also an April in 2021. An attribute should always be functionally dependent on the key. That is basically the same as saying each column has a single value for each value of the key column.

After performing the steps of the first normal form, we got tables with combined keys. The combination as a whole is the key, meaning each attribute should be functionally dependent on each part of the key. Consider, by way of an example, the grade a student gets for an exam. Multiple students do the same exam. Each student could get a different grade. The grade depends on the student. However, this student has to pass multiple exams and can get different grades for each exam. So, the grade *also* depends on the exam. The conclusion is that the grade depends on both the student and the exam. The student and exam together make the combined key, and the grade is the attribute.

Suppose that we used **StudentID** and **ExamID** as key columns. Also, suppose that we have attributes for **Grade** and for **StudentLastname**. The name of a student is in no way dependent on the exams they are taking. This means that the attribute name cannot be part of a table where **ExamID** is part of the key. The attribute should be dependent on the entire key and, with that, on each part of the key, like the grade in this example.

> **Note**
>
> The second normal form states that each attribute (column) is entirely functionally dependent on the entire key.

As with the first normal form, the second normal form solves the problem by splitting the tables with issues into two separate tables. The original table, the one with the original key, will stay. We create a new table with a new key. This key is part of the key of the original table. All columns that depend only on this part should move to the newly created table. All other columns remain in the original table. The result is shown in *Figure 3.5*:

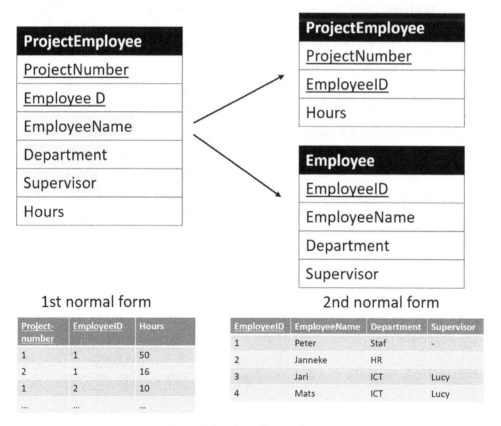

Figure 3.5 – Second normal form

You need to determine for each column in the original table whether it depends solely on **EmployeeID** or on both **EmployeeID** and **ProjectNumber**. A column will never be dependent on just **ProjectNumber**. Columns that depend on **ProjectNumber** alone would have ended up in the other table (with **ProjectNumber** as the key) from the first normal form step.

We need to make a couple of remarks regarding *Figure 3.5*.

Table name

The new table got the name **Employee**. This name logically follows from the key. Each row in this table will describe an employee who is uniquely identified by their **EmployeeID**. That makes **Employee** the most logical name.

Supervisor

The **Supervisor** column is part of the **Employee** table and not part of the **ProjectEmployee** table. Implicitly, we say that the supervisor is somebody in the chain of command. Each employee has one supervisor (boss), like they have one name. Both **EmployeeName** and **Supervisor** are "ordinary" columns in the **Employee** table.

Supervisor could also mean project manager. They supervise the work that is done in the context of the project. That may be a different person to the "everyday" manager. When someone works on multiple projects, that someone may have multiple supervisors.

In the last case, the **Supervisor** column should have ended up in the **Project** table in the first normal form. When each project has a single project manager, like it has a single name, the **Supervisor** attribute should be part of the **Project** table.

Only when there can be multiple project managers per project and each project manager supervises multiple employees does the **Supervisor** column belong to the **ProjectEmployee** table.

Later, we will see what it means that project managers are just employees themselves.

Hours

The **Hours** column stayed behind in the original **ProjectEmployee** table. It should be part of the **Employee** table, when *hours* means the weekly hours as specified in a person's employment contract. It is something we know about each employee, again, like we know an employee's name.

In the **Project Overview** report, the **Hours** column shows how many hours an employee has already worked on the project. The employee called "Peter" worked a different number of hours on project 1 than he did on project 2. Janneke has a different number of hours to Peter. The hours someone worked on a project depends both on the someone and on the project. That is why the **Hours** column should be part of a table with **ProjectNumber** *and* **EmployeeID** making up the key.

Third normal form

Most databases are said to be in the third normal form. This means that the database is normalized until (and including) the third normal form. After the second normal form, there may still be issues in your tables. Take a look at *Figure 3.6*:

EmployeeID	EmployeeName	Department	Supervisor
3	Jari	ICT	Lucy
4	Mats	ICT	Balthazar

Figure 3.6 – Possible issues after the second normal form

Who is the manager of the ICT department? Is it Lucy or is it Balthazar? Or are they both managers in the same department?

We assume here that each department has *one* manager and that the **Supervisor** column on the report is that manager. By now, the word "one" should provide you with a clue regarding the solution. We need a table called **Department** with a column called **Supervisor** or **Manager**. This is exactly what the third normal form is about.

Redundancy is once more the problem. With the assumption about each department having a single manager, we know someone's manager as soon as we know the department in which he or she works. But we store the manager in the **Employee** table. This means that we store the manager's name twice when there are two employees in a department. And as before, when storing a value twice, you run the risk of making a mistake, as in *Figure 3.6*. Lucy is the manager of the department ICT but the second row stores the value Balthazar.

A **Supervisor** column in the **Employee** table does not seem wrong at first sight. That is because in this case, we have a transitive dependency. Whoever is the manager can vary from person to person. However, that is not because we are dealing with another person, but because they work in a different department. Whenever the department is the same, the value for **Supervisor** should also be the same.

> **Note**
>
> The third normal form states that columns need to be directly dependent on the key, and not transitively through another "ordinary" column.

Again, the solution is to split the table that has this issue into two separate tables. You can see the solution in *Figure 3.7*:

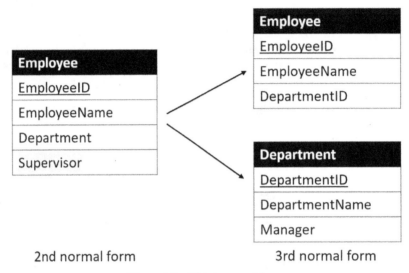

2nd normal form   3rd normal form

Figure 3.7 – Third normal form

We will make a couple of remarks regarding the solution provided in *Figure 3.7*.

Key and name

The conclusion we came to is that **Supervisor** depends on **Department**. The solution is to create a table with **Department** as the key column and make **Supervisor** a column in this newly created table. However, the **Department** column has a text data type and holds a (possibly non-unique) name. With the discussion about surrogate keys from *Chapter 1, Introduction to Databases*, in mind, we chose to introduce a **DepartmentID** column. This will be the efficient surrogate key. We changed the **Department** column to **DepartmentName** to make its content and intention clearer. We also decided that the name **Manager** better describes the functional meaning of the last column than **Supervisor** does.

Using **DepartmentID** as the key, and looking at the data stored in this table, **Department** seems the most logical name for this newly created table.

Foreign key

Notice that the conclusion was that the **Supervisor** column was placed in the wrong table. Nothing was said about the **Department** column. That is why this column stays behind in the original table. In that table, it became a foreign key to the newly created table. Do notice that we had to change it to **DepartmentID** to be this foreign key.

Boyce-Codd and the fourth normal form

In most cases, you are done after the third normal form. You designed tables that store the data needed for the original report without any redundancy left. In rare cases, however, there may still be some issues left in the tables that constitute the third normal form.

Boyce-Codd normal form

After the third normal form, you may end up with composite keys that are made up of overlapping candidate keys. That is not the case in our example, so we are done. However, for the sake of completeness, let's have a look at what overlapping candidate keys means. The Boyce-Codd normal form deals with this issue.

Have a look at *Figure 3.8*, showing an example of the Boyce-Codd normal form. A student is enlisted in multiple classes. A student has one teacher for each subject/class. The combination of student and teacher is a candidate key (where we assume that a teacher only teaches one subject) in the table as shown in *Figure 3.8*. The combination of student and class is also a candidate key. Both combinations hold unique values. The **Student** column is known as an overlapping key:

Class
Student
Class
Teacher

Student	Class	Teacher
1	English	MyLady Engels
1	French	Frans Docent
2	English	MyLady Engels
3	Physics	Mister Einstein
4	English	Mister Taal

Figure 3.8 – Possible issues after the third normal form

In the preceding example, there is a dependency between teacher and class. When we keep the table we have in *Figure 3.8* as is, there will be an input anomaly. Suppose you hire a new teacher. As long as no student is assigned to this new teacher, we cannot add them to the database. The key is always obligatory, and we need a student to create a key.

As always, the solution is to split the table and create a separate **Teacher** table to store the dependency that class has on the teacher. A teacher can be linked to the class they teach and a student can be linked to a teacher and then, through the teacher, to a class. See *Figure 3.9*:

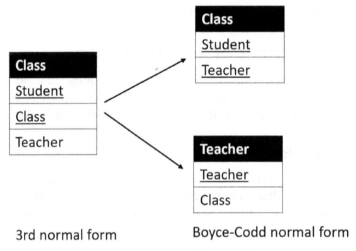

Figure 3.9 – Boyce-Codd normal form

Notice that we didn't use surrogate keys in this example. That is just for simplicity. Normally you would.

Fourth normal form

After the Boyce-Codd normal form, you may need to look at the fourth normal form. The fourth normal form is about multi-valued dependencies. These may arise when three or more columns together form the primary key, with two of the columns being independent of each other. Take a look at *Figure 3.10* to see an example:

Employee
Employee
Skill
Hobby

Employee	Skill	Hobby
1	T-SQL	Cycling
1	Scrum	Gaming
2	Data modelling	Cycling
2	C#	Gaming

Figure 3.10 – Possible issues after the Boyce-Codd normal form

In the example of *Figure 3.10*, we assume that **Skill** and **Hobby** are independent of one another. This means that the combinations shown in the example are completely random. Employee 1 has two skills and two hobbies. Combining each skill with each hobby would create four rows for employee 1. The same goes for employee 2, meaning we should have eight rows instead of four. The four rows we have are a random subset of what we would have were the table complete. Can we live with randomness like this in our database? And what if an employee has skills but no hobbies? With **Hobby** as part of the key, we would not be able to create a row. Both a skill and a hobby are required.

We need to split the table into two separate tables, as demonstrated in *Figure 3.11*:

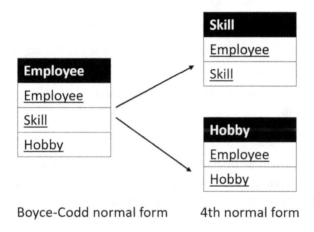

Figure 3.11 – Fourth normal form

Both the Boyce-Codd and the fourth normal form are rare. After completing the steps that lead to the third normal form, you are usually done. But do keep an eye out for them. Be aware of tables that have multiple columns in the primary key. Thinking logically is usually enough to spot the problem.

Normalizing – a recap

In most cases, databases are normalized until the third normal form. They are said to be in the third normal form. Each step tackles a potential redundancy problem. The steps focus on the following:

- The preliminary step (0th normal form) recognizes the columns that we need to work with:

  a. Don't store values that can be calculated from other values (process data).

  b. Don't store parameters (values coming from a place (system) other than the database.

- The first normal form ensures that all the columns store atomic values and that they are all unique:

  Do not store a comma-separated list of values in a column. For example, there should be no **Hobby** column with the value "Cycling, gaming, studying".

  No columns such as **Hobby1**, **Hobby2**, and **Hobby3**.

- The second normal form is all about functional dependencies:

  Each column of a table should have a functional dependency on the entire key.

  When part of the key becomes different, the column itself can be different as well.

- The third normal form tackles transitive dependencies:

  A column should have a direct functional dependency in the key and not indirectly through another column.

- The Boyce-Codd normal form removes tables where a key column depends on an "ordinary" column.

- The fourth normal form resolves issues arising from multi-valued dependencies.

Every step associated with normalizing splits a table with issues into two tables. With the first, second, and third normal forms, you end up with the original table and a new table with a new key. Both tables that remain after the split always have a relationship with one another. That relationship is always a primary key–foreign key relationship.

An important element when normalizing data is to get the details right. A **Name** column may be too vague. The name of a person is something entirely different to the name of a product. The name of a person may need to be split into parts such as a first name and last name.

The steps described in this paragraph may be formal and abstract. It is probably mathematical set-based theory that is behind all this. On the other side, it is all logic. Think logically. Create some example rows for yourself to see what the tables you create mean. Pay extra intention to combined keys and what they mean.

The end result of us normalizing the **Project Overview** report is four different tables: **Project**, **ProjectEmployee**, **Employee**, and **Department**, as shown in *Figure 3.12*:

Project	ProjectEmployee	Employee	Department
ProjectNumber	ProjectNumber	EmployeeID	DepartmentID
ProjectName	EmployeeID	EmployeeName	DepartmentName
Budget	Hours	DepartmentID	Manager

Figure 3.12 – Resulting tables

An alternative approach to normalizing data

In the previous section, we normalized one report. In real life, there will be a multitude of reports, screens, and forms to analyze. A lot of them will use the same columns. To create a complete database schema, we need to analyze and normalize all of them. For now, let's assume that there is a second report, **Project Progress**, that we need to normalize. You can see the report in *Figure 3.13*:

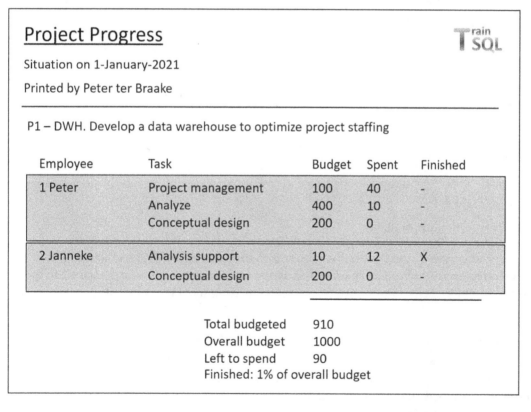

Figure 3.13 – Project progress

We could go through all the formal steps for this report as we have done for the first report. Instead, we will use an alternative approach. This method sort of involves combining entity analysis and normalizing data. In real life, experienced database designers do not go through all the formal steps. Some even say that we no longer normalize nowadays, but not doing the formal steps explicitly doesn't mean you are not applying the underlying theory.

Step 1

The first step to undertake is the same as with normalizing. We need a list of columns storing all the values we see on the report. You need to remove process data (values that can be computed) and parameters (values not coming from the database). The **Project Progress** report of *Figure 3.12* leads to the following list of columns:

- **ProjectNumber**
- **ProjectName**
- **ProjectDescription**
- **EmployeeID**
- **EmployeeName**
- **Task**
- **HoursBudgeted**
- **HoursSpent**
- **Finished**
- **OverallHoursBudgeted**

Notice that we have already split the project into **ProjectNumber**, **ProjectName**, and **ProjectDescription**. We also changed the names of the columns slightly compared to the headers used in the report in an effort to make the meaning of the columns more intuitive.

Step 2

In the next step, we try to recognize entities from this list. The column that makes you recognize an entity can (for now) be seen as its key, so we underline that column. In our list, we immediately recognize the **ProjectNumber** and **EmployeeID** columns as (possible) keys of entities.

Upon analyzing the report, we see gray sections that are similar in structure. The values in these sections are different, but structurally speaking, they are identical. Within a section, we see multiple lines. Each line holds the same type of information (the same columns), but with different values. Each section relates to an employee. Each line describes a task being performed within the project. This makes both **Employee** and **Task** entities. We already have an entity, **Employee**, because we recognized **EmployeeID** as a key, and we found a new entity called **Task**.

Notice that we are just performing the first step of normalizing data, the first normal form.

With the (possible) keys underlined, our list now looks like this:

- **ProjectNumber**
- **ProjectName**
- **ProjectDescription**
- **EmployeeID**
- **EmployeeName**
- **Task**
- **HoursBudgeted**
- **HoursSpent**
- **Finished**
- **OverallHoursBudgeted**

Step 3

The next step is to create a table for each key you marked in the list of columns. For all the other columns, we try to determine which of the keys it belongs to (depends on). In other words, we are looking for functional dependencies. This step leads to the tables in *Figure 3.14*:

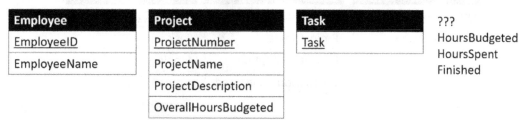

Figure 3.14 – Tables after step 3

The **HoursBudgeted**, **HoursSpent**, and **Finished** columns are written outside the tables with question marks. They do not belong to any of the three tables. For instance, 100 hours are budgeted for project management. Does this make **HoursBudgeted** a column of the **Task** entity?

When, for instance, the "Project management" task description is unique for this project and is being carried out by just one employee, then the hours budgeted for this task depend solely on the task. When there are more projects with a "Project management" task, the hours budgeted depend on the task *and* on the project. Bigger projects probably have more hours budgeted for this task than smaller projects. There clearly exists a dependency on the project.

We need to investigate **Task** a little more closely. Is a task uniquely assigned to one person? In that case, the **EmployeeID** column is a column in the table where **Task** is the key. When a single task can be assigned to multiple people, each of these persons may have a different number of hours assigned to spend on this task. That makes **HoursBudgeted** dependent on both the task *and* the employee, which makes the combination of the two a key.

As we can see in the report shown in *Figure 3.12*, the "Conceptual design" task is carried out by both Peter and Janneke. We assume that another project can also have a task called "Conceptual design." We also know that a project consists of multiple tasks. All this together means that we need a combined key with the columns **ProjectNumber**, **EmployeeID**, and **Task**.

Step 4

In this step, we create tables with composite keys to assign the columns written with question marks in the previous steps to tables. We also need to make sure that all the tables have relationships using foreign keys.

Using everything we have learned so far, we can now create the tables in *Figure 3.15*:

Employee	Project	Task
EmployeeID	ProjectNumber	ProjectNumber
EmployeeName	ProjectName	Task
	ProjectDescription	EmployeeID
	OverallHoursBudgeted	HoursBudgeted
		HoursSpent
		Finished

Figure 3.15 – Tables after step 4

To finish, you need to check that everything makes sense. This check is a combination of what you learned in entity analysis and what you have learned so far in this chapter. The alternative approach just described may seem easier, quicker, and more intuitive than the formal steps described in the previous section. The risk of making mistakes is probably higher. The end result should be the same with both approaches. It leads to a set of related tables that adhere to the third normal form.

Integrating separate results

We analyzed two different reports in the two previous sections of this chapter. Both reports provide information on the same project organization and show data coming from the same database. The results of the independently analyzed reports need to be integrated into a single database schema.

During integration, we merge the result of normalizing the information requirements. We need to pay attention to the following:

- Tables that describe the same entity

- Homonyms and synonyms

- Process data

Tables that are the same

An entity (a table) is defined by its key. So, we first look for tables with the same key. We can merge these tables.

Both in *Figure 3.7* and in *Figure 3.15*, we see a table called **Project**. The fact that we chose the same name is irrelevant, just a coincidence. The fact that both tables have the same key, **ProjectNumber**, means that they are actually the same table. You merge them by creating a table that contains all the columns from both tables. Do not create duplicate columns. Each column, whether in the first table or in the second table or in both tables, should appear in the new table just once.

Homonyms and synonyms

When merging the tables called **Project**, we see a column called **Budget** in the first table and a column called **OverallHoursBudgeted** in the second table. It could be that these columns are indeed different because they have different names. One might be the budget in euros, whereas the other is obviously a budget in hours. They could also be the same piece of data that has a different label in different reports. In that case, they are synonyms: a different name but the same meaning.

> **Note**
> Synonyms are where two columns have different names but store the same
> data, in which case the two columns are actually just one column.

In our example, the columns are identical. We use only one of them in the end result.

We have another synonym in our example. The **Department** table has a **Manager** column.
The word "Manager" describes the role of this column within this table. However, a
manager is an employee. We missed that during the normalization phase (or perhaps you
didn't). At this stage, we need to recognize that the **Manager** column holds an employee.
This raises a question: do we store the name of the manager in the **Employee** table or their
EmployeeID? The last option is more efficient and less redundant. The **Manager** column
now becomes a foreign key to the **Employee** table.

Homonyms are when two columns have the same name but store different pieces of
information. In the example, we could have two columns called **Name**, the name of an
employee and the name of a project. It might be a good idea to change the generic term
"Name" to something more descriptive, as we have already done in our example so far.

> **Note**
> Homonyms are when two columns with the same name actually store different
> information.

Other really common examples of homonyms are date and price. Does price mean the
purchase price, sales price, or quoted price? Is the price inclusive or exclusive of VAT?
The way you use a column called **Price**, the table it is in, meaning the key it depends on,
should clarify this somewhat already. A list price is probably part of a **Product** table,
whereas a sales price is a column in the **Orderline** table.

Try to avoid using generic terms such as date and price as column names. Make your
column names more descriptive.

Process data

We excluded process data from the preliminary step of normalizing data. During
integration, we may reintroduce process data. We need to recognize that and delete the
process data.

Look, for example, at the **ProjectEmployee** table in *Figure 3.7*, and the **Task** table in *Figure 3.14*. Both have a different key, meaning they are different tables. The **ProjectEmployee** table has a column called **Hours**. Looking at project 1 in the **Project Overview** report, we can see the value 50 for the employee called Peter. Looking at the **Project Progress** report, we can add together all the values in the **HoursSpent** column for Peter. Peter spends 40 hours on project management and 10 hours on analysis. Combined, he spends 50 hours working on project 1!

Apparently, the **Hours** column in the **Project Overview** report is the number of hours spent, and not the number of hours budgeted. Because the second report shows more details by listing individual tasks, the **Hours** column in the **ProjectEmployee** table becomes process data. It can be calculated by adding the relevant rows from the **Task** table. That means we need to remove **Hours** from the **ProjectEmployee** table.

Removing the **Hours** column means that our **ProjectEmployee** table now consists of a key without "normal" columns. By itself, this is OK. It still holds information: which employees work on which projects.

Looking closer at the **Task** table reveals that the same information can be derived from **Task** as well. **Task** holds information on all the tasks that are part of a project. We also know for each task who it is assigned to. Through this, we know all the employees participating in the project. The **Task** table holds the same information, just at a more detailed level. We can remove the entire **ProjectEmployee** table because this constitutes process data in its entirety.

Integration seems easy. It can be a difficult step because it is so easy to miss synonyms, homonyms, and process data. You need to know the data and the process very well in order to perform this step well.

Now that we are done normalizing our data and integrating the result, it is time for the final step: putting the results in an ERD.

Entity relationship diagram

The final step is to draw the ERD, just like we did in *Chapter 2, Entity Analysis*. The understanding of ERDs you acquired in the aforementioned chapter will serve you well in understanding the results of normalizing. The key part of this step is to recognize all foreign keys. Using entity analysis, you *think* entities have relationships. Recognizing foreign keys means you now *know* that there are relationships and that you can tell the database about them later when implementing the database. You can show all columns in an ERD, or just show the tables as rectangles with just the table name written in them.

The ERD of our project administration is now as can be seen in *Figure 3.16*:

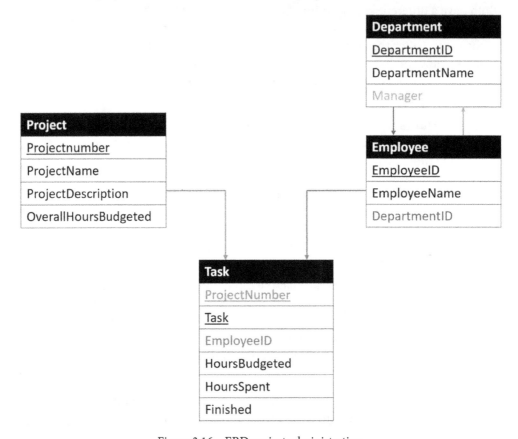

Figure 3.16 – ERD project administration

In *Figure 3.16*, the colors indicate the foreign keys and the relationships they implement. The value of drawing the ERD is twofold:

- To get an overall picture
- To show the necessary details

The ERD provides an overall picture

Using *Figure 3.14*, you know each table and exactly what each table looks like. The relationships between the tables are not immediately evident from *Figure 3.14*. By drawing the ERD, you don't add new information to the design. You do, however, create extra insight on account of the overview that an ERD provides. It can also serve really well as database documentation. Both application and BI developers will benefit from being able to use the ERD.

Often, the most important part of drawing the ERD is the extra check it provides you with. Sometimes, the ERD "doesn't feel right." Plain logic should provide some leads as to what might not be right. Going back to the details of normalizing will lead you to solutions.

The ERD shows the necessary details

Performing entity analysis mainly involves common sense. However, do the relationships you draw really exist? Can you tell the database about the relationship so that it can provide referential integrity?

Figure 3.15 shows foreign keys and the relationship they implement in matching colors. You always need matching primary key and foreign key pairs to create a relationship. When you want a relationship between two tables, you simply look for the foreign key. If there is no foreign key, there is no relationship. You may be able to create a foreign key if you really decide you need a relationship.

There are a couple of additional remarks to be made regarding *Figure 3.15*. For starters, note that we use arrows, with the arrowhead pointing to the many side of the relationship. Also, notice that the **Employee** and **Project** entities have a many-to-many relationship. Many-to-many relationships need to be replaced by bridge tables. This happened naturally during normalization and resulted in the **Task** table.

Also note that we have two relationships between the **Department** and **Employee** entities. The direction of the relationship is reversed. The left (purple) relationship comes from the **DepartmentID** foreign key in the **Employee** table. This relationship describes the department in which an employee works. The right (orange) relationship is based on the **Manager** foreign key in the **Department** table. It describes which employee manages the department. Both relationships are very different and can co-exist.

Summary

Normalizing data tells us how to store all the data we need in order to provide people and applications with the information they need. For each individual column, we know exactly the table to create to store it in. We removed all the redundancy from the database. Removing redundancy optimizes the storage we need to implement the database. It also optimizes writes in the database, while minimizing the risk of inconsistencies. In other words, we created a database optimized for an OLTP workload.

There are some additional steps to perform before we can actually implement the database in Azure. You will learn about these steps in *Chapter 4, Provisioning and Implementing an Azure SQL DB*. In that chapter, you also learned how to implement the database as an Azure SQL database.

Exercises

Exercise 1 – Stock management of a bicycle shop

A bicycle shop specializing in high-end road bikes has trouble keeping their stock supplies at acceptable levels. We therefore created a new database to help them get a better insight into the products they have in stock.

The shop has a list of suppliers. Many products can be bought through different suppliers. In this process, it is important to know which list price each supplier is using and what the delivery time is. Based on that information, a purchaser orders new products:

- Draw an ERD based on the description provided. Use your entity analysis skills, as acquired in *Chapter 2, Entity Analysis*. You will normalize the data in later steps in this exercise.

- Normalize each report as shown in *Figures 3.17*, *3.18*, and *3.19* individually.

Figure 3.17 shows a paper-based purchase order:

<table>
<tr><td colspan="3">PURCHASE ORDER</td><td>11 January 2021</td></tr>
<tr><td>Purchase order number:</td><td>12345</td><td></td><td>Purchaser: JPD</td></tr>
<tr><td>Supplier:</td><td>Peter's wholesale</td><td></td><td></td></tr>
<tr><td>Address</td><td>Quickbike Alley 1a</td><td></td><td></td></tr>
<tr><td></td><td>9876 BikeCity</td><td></td><td></td></tr>
</table>

Article	Description	Quantity	Price
A5	Handlebar tape	100	76,-
O8	Alu spokes	250	25,-
A-98	Bidon	20	70,-
	Total		Euro 171,-

Delivery date: 26 January 2021

Figure 3.17 – Purchase order

Figure 3.18 shows a warehouse card that can be printed for each product:

WAREHOUSE CARD		11 January 2021

Article number: A-98   Warehouse location:

Description: Bidon with logo   Rack: 5

Current stock: 3   Area: A

Bidons still on order:

Purchase order number:	Delivery date:	Quantity:
12345	26 January 2021	20

Figure 3.18 – Warehouse card

Figure 3.19 shows a list of products that a supplier can deliver. The shop has such a list for all their suppliers:

Peter's wholesale - Catalog

Supplier: 5

Name: Peter's wholesale

Address: Quickbike Alley 1a

                9876 BikeCity

Products:

Product number:	Article:	Description:	Delivery time:
1	A-98	Bidon with logo	1 week
2	O8	Alu spokes	3 weeks
3	A5	Handlebar tape, yellow	2 weeks

Figure 3.19 – Catalog

- Draw an integrated ERD describing the entire database.

  How does your ERD compare to the one you drew in *step 1*?

4
Provisioning and Implementing an Azure SQL DB

After creating a conceptual model using normalization techniques (*Chapter 3, Normalizing Data*), it is time to actually create a database. We first need to transform the conceptual design into a physical design. Some details are still missing. You will learn about all these (technical) details required to implement a database as an Azure SQL database in this chapter. In the second part of the chapter, we will actually provision the database and create the designed schema.

In this chapter, you will learn about the following:

- SQL Server data types and their relevance
- Quantifying the data model
- Provisioning an Azure SQL database
- Implementing tables and constraints using data definition language

Technical requirements

You will need the following if you wish to successfully complete this chapter:

- A connection to the internet and a modern browser

- An Azure subscription with permission to create new services

- Permission and the ability to download and install Azure Data Studio

Understanding SQL Server data types

The conceptual data model we have created thus far defines what tables a database consists of. Also, it tells you all the columns that form each table. For instance, a table called `Product` may contain the `ProductName`, `ListPrice`, `Category`, and `EndDate` columns. Before we can actually create this table, we need to know what sort of values each column will contain. In other words, we need to choose an appropriate data type. The data type of a column determines three things:

- Which types of values can be stored in the column. Can you store numbers, such as product prices, or dates, such as transaction dates?

- What computations or manipulations you can do with the data. For example, you can add two numbers together, but you cannot multiply two dates.

- The efficiency of both data storage and data manipulations. Some data types are smaller than others, making them more efficient to store and work with.

Choosing the proper data type is a really important part of creating a database. I once had a table that used 800 GB of storage. By choosing more efficient data types, the same data could have been stored in just 600 GB. That saves 25% on storage. It also saves a lot in terms of processing power when manipulating this data.

> **Note**
> More efficient data types make an Azure SQL database cheaper while improving performance at the same time.

We can roughly distinguish between three types of data:

- Numerical data

- Alphanumerical data

- Dates

Let's look at each of these types in detail in the following sections.

Numerical data

Numbers are numerical data. You can perform calculations with numerical data. You don't necessarily have to do calculations, but you can. **Numerical data** is the most native type of data for computers, which makes it the most efficient data type. In the previously mentioned example of a `Product` table, the `ListPrice` column is a clear example of a column that should have a numerical data type.

We can divide numerical data into two separate data types:

- Whole numbers
- Numbers with a decimal point

Whole numbers

Whole numbers are integer values or, in short, just integers. Integer values *do not* have digits after the decimal point. SQL Server has four different data types for whole numbers. These are as follows:

- `TINYINT` (stores values from 0 to 255)
- `SMALLINT` (stores values from -32,768 to +32,767)
- `INT` (stores values from -2,147,483,648 to 2,147,483,647)
- `BIGINT` (stores values from -9,223,372,036,854,775,808 to 9,223,372,036,854,775,807)

What all four have in common is that they can only store whole numbers. They are 1 byte, 2 bytes, 4 bytes, and 8 bytes in size, respectively. That means that it takes 4 bytes to store a single value of the `INT` data type. It also takes 4 bytes of memory to hold that same value in memory. The difference in size translates into a difference in how big the largest value is that can be stored. In a column of the `TINYINT` data type, you can store values ranging from 0 (zero) to 255. The range of possible values increases with the size of the data type. You can see the range in the previous bulleted list.

If you select a data type that is too small, you will get into trouble. Suppose, for instance, you choose `TINYINT` as the data type for a `ProductID` column that is the primary key of the `Product` table. Since the primary key holds unique values, you can now store a maximum of 255 rows in the table. This means that your table can only store 255 different products.

If, on the other hand, you opt for BIGINT as the data type, you waste resources, such as storage and memory. Suppose you have a factSales table with a foreign key, ProductID. Also, suppose that this table has a billion rows. The difference in storage between the ProductID column having the INT data type compared with the BIGINT data type equals 4 times a billion bytes. That is a little less than 4 GB. If you always choose BIGINT over INT to be on the safe side, with the range of values that you will be able to store, you will waste a lot of memory and storage. You will want to choose the smallest data type that can guarantee storing all the values you have to date and will ever have in the future. When in doubt, you're better off choosing a bigger data type over a smaller data type.

Numbers with a decimal point

Next to whole numbers, we have numbers that have digits after the decimal point. The ListPrice column from our Product table is again a good example. The price of a product could very well be something such as €3.99. We can divide data types that are able to store values with a decimal point into two categories:

- Exact numeric
- Approximate numeric

Exact numeric

SQL Server has NUMERIC and DECIMAL as exact numeric data types that store values with a decimal point. NUMERIC and DECIMAL are equivalent in usage.

> **Note**
>
> From this point on, we will use NUMERIC in this book.

When you create a column with the NUMERIC data type, you need to specify the number of digits that you require (known as *precision*) and the number of digits you need after the decimal point (known as *scale*). You could, for instance, create a column specifying the data type to be NUMERIC (5, 2). This means that you have five digits in total, two of which are after the decimal point. This, in turn, means that the smallest value you can store in this column is -999.99, while the biggest value is 999.99.

The biggest value you can use for precision is 38. The size of a column with the INT data type depends on the precision you choose. You can look this up in the SQL Server documentation on docs.microsoft.com.

SQL Server also has the MONEY and SMALLMONEY data types. These are functionally the same as numeric. Both store four digits after the decimal point. They take 8 and 4 bytes of storage, respectively. MONEY can store values up to 922,337,203,685,477.5807, whereas SMALLMONEY can only store values up to 214,748.3647.

Approximate numeric

You may need to store bigger values than NUMERIC can hold, or you may need to store values with more digits after the decimal point than NUMERIC is capable of doing. In that case, you can use the FLOAT or REAL data type. The REAL data type takes 4 bytes of storage and stores values consisting of 38 digits. The decimal point can be anywhere between those digits. When the decimal point is at the start, you have a value with 38 digits after the decimal point.

The FLOAT data type is either 4 or 8 bytes in size. You can specify the mantissa of the stored value when stored, as in the scientific notation. You can find the exact range of values in the official SQL Server documentation (docs.microsoft.com/en-us/sql/t-sql/data-types/float-and-real-transact-sql).

The drawback of FLOAT and REAL is that both data types store approximate values. Have a look at *Figure 4.1*:

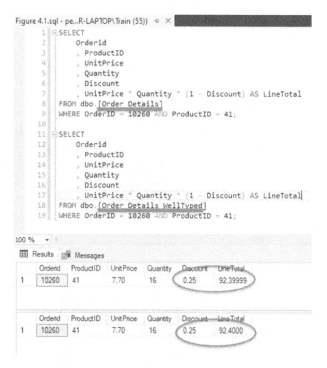

Figure 4.1 – Approximate numeric data types

Figure 4.1 shows two identical queries. The only difference is that the first query uses the [**Order Details**] table where the **Discount** column is of the REAL data type. The second query uses the [**Order Details WellTyped**] table, which is identical to [**Order Details**], except that the **Discount** column is now of the NUMERIC(10, 2) data type. The last column is a computed column where a transaction amount is calculated. Even though the values of the real columns look identical, the result of the calculation is not. Because REAL is an approximate data type, you get approximation errors when REAL is used in calculations.

> **Note**
>
> Try to avoid approximate numeric data types.

The generic rule is that you can store bigger values when you use more memory. Using more memory per stored value might lead to performance bottlenecks sooner than using less memory. With small databases, this is hardly relevant, but with bigger databases, meaning you store more rows per table, this can have a huge impact. Therefore, make a well-informed choice based on what you just learned about the data types and what you know about the actual values that your database needs to store. When in doubt, choosing the bigger data type is the safer choice to make.

Alphanumerical data

The Product example table from the start of this section has the ProductName, ListPrice, Category, and EndDate columns.

Of course, a product name is not a number and therefore cannot be stored in a numerical data type. We need an alphanumerical data type. Columns with an alphanumerical data type can store anything you can type on your keyboard. It is also called *character data*, or simply a string. A product name might be something such as *Peanut butter*, which contains letters. This is an example of alphanumerical data. Besides letters, it may contain special characters, such as question marks, exclamation marks, or the @ sign. It may also contain numbers. Often, this data type is referred to as *text*.

You could store numbers as text as well. But there are two disadvantages:

- Text data takes more storage and memory than numerical data.

- You cannot do any calculations on text data. You can change text data that only contains digits into numerical data and then start doing calculations. However, doing so will cost you in terms of performance.

Having said that, sometimes, numbers are more alphanumerical in nature than numerical. Take, for instance, a house number. This suggests it is a number. However, it is unlikely that you need to calculate with these numbers. Besides, you may have a house number such as 23A, which you cannot store when you make a `HouseNumber` column numerical. The `HouseNumber` column should have a text data type.

There are two considerations to bear in mind when using alphanumerical data:

- Are different alphabets being used for your data?

- Do you need your data to be case-sensitive?

Are different alphabets being used for your data? SQL Server has four different alphanumerical data types. These are as follows:

- CHAR

- VARCHAR

- NCHAR

- NVARCHAR

Notice that the last two data types listed start with the letter *N*. This letter stands for the *National* character set. These data types use two bytes of memory to store a single character, whereas CHAR and VARCHAR only use a single byte per character. So, CHAR and VARCHAR are more efficient than NCHAR and NVARCHAR. But that comes with a price. You need to specify the character set or, simply put, the alphabet that you will use.

Suppose you create an `Employee` table with a `LastName` column. You will, sooner or later, have people working for you with names originating from foreign countries. They might have letters in their name that are not part of your alphabet. Using CHAR and VARCHAR as the data type for the `LastName` column means you cannot store their names correctly.

NCHAR and NVARCHAR use more storage than CHAR and VARCHAR, but are independent of the chosen alphabet, meaning you can store Chinese, Hebrew, and Greek characters in them, or characters from any other alphabet.

In SQL Server, you choose the character set (alphabet) to use by specifying the **collation**. You can do that when you create a new database. The specified collation will serve as the default for all the columns you create. Whenever you create new columns in tables using an alphanumerical data type, you can specify the collation to use for that specific column. The best practice is to use the same collation for all columns when possible.

The default collation of an Azure SQL database is `SQL_Latin1_General_CP1_CI_AS`. The `Latin1` part specifies that you will use the Latin alphabet, which contains the 26 letters used in English. You will see collation later in this chapter when we provision an Azure SQL database.

Do you need your data to be case-sensitive?

The second consideration when using text data is whether or not that text needs to be case-sensitive. Is *PEANUT BUTTER* the same as *Peanut butter*? When you create a unique constraint on a `ProductName` column and the column is case-sensitive, both example values are actually different and will be allowed in the column. When the column is not case-sensitive, both values are considered to be identical.

You may need to create a new database for an existing application. You need to know whether the logic in the application depends on case sensitivity. Looking back at the previously mentioned default collation (`SQL_Latin1_General_CP1_CI_AS`), the `CI` part stands for *case-insensitive*, so a small *a* and a capital *A* are considered to be the same. You can change the `CI` into `CS` to make the database case-sensitive.

The `AS` in the collation specifies that the database is accent-sensitive. An *a* and an *á* are different. You can change this by using `AI` instead of `AS`.

For the sake of completeness, `Latin1` specifies the English alphabet, but there are a lot of variations on the English alphabet. The `General_CP1` part makes it specific.

Depending on where you are from, you may want to choose a different collation for your database. By way of a recap, `NCHAR` and `NVARCHAR` are independent of the alphabet part of the collation. Whether text is case-sensitive or not affects all alphanumerical data types.

Varying-length data types

We discussed the difference between `NCHAR` and `NVARCHAR` on the one hand and `CHAR` and `VARCHAR` on the other. Now we need to focus on the difference between `CHAR` and `NCHAR` on the one hand and `VARCHAR` and `NVARCHAR` on the other. In other words, what does the `VAR` part stand for?

When you create an alphanumerical column, you need to specify the length of that column. A column definition would look something like this:

```
ProductName   NCHAR(50)
```

Or it may look like this:

```
ProductName   NVARCHAR(50)
```

With the NCHAR(50) data type, the number 50 is the length of the string stored. The length is always 50, even if you store the name Cheese, which has only 6 characters in it. This means that independent of the actual data stored, enough space is reserved to store 50 characters, which, in the case of NCHAR(50), means 100 bytes. With Cheese as the name, you waste 88 bytes of storage.

With the NVARCHAR(50) data type, the number 50 is the maximum length of the string stored. But when you store Cheese, this will use 6 times 2 bytes for the 6 characters in the name Cheese, plus 2 extra bytes to store the actual length of the string stored. Therefore, our Cheese value now only uses 14 bytes. We save 86 bytes compared to the NCHAR(50) data type.

> **Note**
>
> The phrase VAR stands for **varying in length**.

Data types that can vary in length are slightly less efficient than fixed-length data types. So, when you need to store US state codes, which are always two letters coming from the Latin alphabet, use CHAR(2) as the data type. However, the inefficiency of data types of varying length has a far smaller impact than the storage and **I/O (input/output** to actual files on storage) savings with names that are inherently different in length. So, a varying-length data type is, in most cases, the best choice.

The maximum length of a string cannot exceed 8,000 bytes. That makes 4,000 the maximum you can specify for NCHAR and NVARCHAR because you specify the number of characters, not the number of bytes used. If you require larger values, you can use the MAX keyword. A column of the NVARCHAR(MAX) data type can store up to 2 GB of text data.

Dates

The last of the three main data types that we distinguished at the start of this section is **dates**. The Product table from the example at the beginning of this section has an EndDate column. This column specifies when we stopped selling a product. There are two reasons for storing this value as a DATE data type:

- Data quality
- Functionality

Data quality

Sometimes, dates get stored in columns with alphanumerical data types. Alphanumerical data types are less efficient than DATE data types. But even more importantly, you can store anything in text, including a value such as 29-feb-2022. This, however, is only a valid date if 2022 is a leap year, which it is not. Trying to store the value 29-feb-2022 in a column of the DATE data type will raise an error; storing it as text is just fine.

Functionality

As with numerical data, you need to be able to do calculations with dates. For instance, you might need to calculate the number of days between an order date and the delivery date. Or you might need to know in which week a particular date falls.

Both these examples use special date functions. SQL Server has a lot of special functions allowing you to do calculations on dates. All these functions use values that are stored as a date and not values stored as text.

As with numerical and alphanumerical data types, SQL Server has a few different DATE data types. These are as follows:

- DATE
- TIME
- DATETIME2/DATETIME/SMALLDATETIME
- DATETIMEOFFSET

As you can see, there are special data types for date, time, and datetime. The DATE data type only stores dates. It stores dates in the range of 1-Jan-0000 until 31-Dec-9999 and uses just 3 bytes of storage. It doesn't include time. In most cases where we store the birth dates of people, we don't care about what time of day they were born. Using DATE is then the best choice: you don't waste storage on storing a time part that is not relevant. Plus, you don't complicate queries because they don't have to take into account the fact that there might be a time aspect to the date value.

In other cases, however, the exact point in time during a day may be of interest. You now have two options. You either store the date part and the time part separately, or you combine them into a single value of the DATETIME2 data type.

DATETIME2 stores the date and time as one value. Whenever you create a column of the DATETIME2 data type, you need to specify the precision. If you create a column as DATETIME2(7), you can store up to seven decimal places of the second. This means the precision is 100 nanoseconds. With just DATETIME2, the precision is seconds. DATETIME2 takes 6, 7, or 8 bytes of storage, depending on the precision you choose. When you use date and time as a whole in the majority of your queries, storing it as a whole makes the most sense.

If you use the time part mainly independently of the date itself, it makes sense to store the date and time separately. For instance, when you do a lot of analysis in terms of when, during the day, the workload increases to optimize resource utilization, you may want to work with just time. Time takes 5 bytes of storage. As with DATETIME2, you can specify the precision.

As you can see in the preceding bulleted list, SQL Server also uses DATETIME/ SMALLDATETIME. Both are older data types and I recommend that you use the newer ones. The newer data types are more efficient and have better precision.

The last data type shown in the list is DATETIMEOFFSET. This data type allows you to store an extra offset that specifies how much that datetime differs from UTC. This might be relevant when you need to compare dates coming from different parts of the world. DATETIMEOFFSET takes 10 bytes of storage.

Again, as with numerical and alphanumerical data types, you first make sure that you understand the values that you need to store. You then make sure you know what kind of manipulations you need to perform on the data. Lastly, you choose the most efficient data type (the smallest in size) that provides all the functionality you need.

Other data types

Numerical, alphanumerical, and dates are the main data types we work with in SQL Server. However, there are several other data types as well.

Binary

A computer basically works with 0s and 1s. This is called binary. Especially when using text data types, the 0s and 1s need to be translated into letters, digits, and special characters. It does not always make sense to translate it into something more meaningful for people, for instance, when you store encrypted data or hashed data. Another example could be storing images or sound bites. In these scenarios, the best data type is binary.

SQL Server has two data types for storing binary data:

- BINARY
- VARBINARY

As with character data, you specify the length of your data after the data type, for example, VARBINARY(50). The number is either the length or the maximum length of your data in bytes depending on the addition of VAR.

XML/JSON

SQL Server has a special data type for storing XML data. SQL Server even supports XQuery to write queries against data stored as XML. SQL Server has no special data type for JSON data. You can store JSON in "normal" alphanumerical data types. SQL Server does have functions to parse both XML and JSON data. It can also return result sets of queries in the form of either XML or JSON.

We will talk more about XML and JSON in *Chapter 5*, *Designing a NoSQL Database*.

GUID

GUID stands for **globally unique identifier**. In some applications, having a key that is unique within the table is not enough. You need values to be unique across the world regardless of the table they are in. You may, for instance, build a distributed solution with an Order table in a database hosted in the US, and also an Order table in a database hosted in Europe. Suppose you need the OrderID column to have unique values over both tables. Also, you want to keep in mind that you might create a database with an Order table in Asia in the future as well.

Using IDENTITY as a way to generate unique values now doesn't work. It generates unique values within a single table. You could use the NEWID() function to generate guaranteed GUIDs and use the GUID data type for the primary key column.

Make sure that you absolutely need your values to be globally unique. GUID gives you that, but GUIDs are not the most efficient data type, especially not for primary keys. It is not a recommended best practice to use GUIDs as primary keys.

Spatial data

The `Geography` and `Geometry` data types allow you to store coordinates such as GPS coordinates or longitude and latitude. They also allow you to create and store shapes, such as, for instance, a map of your office building. If you have this kind of data, you can now do all sorts of manipulations on the data. You could, for instance, write a query that retrieves all rooms for which the nearest coffee machine is more than 50 meters away.

The `Geography` data type considers the earth to be a sphere. With `Geometry`, you create shapes and coordinates on a flat surface.

Refer to the official documentation to see the full spectrum of possibilities of these spatial data types (`docs.microsoft.com/en-us/sql/relational-databases/spatial/spatial-data-sql-server?view=sql-server-ver15`).

Additional data types

SQL Server has even more data types. To get the full list, please visit the official documentation on `docs.microsoft.com/en-us/sql/t-sql/data-types/data-types-transact-sql?view=sql-server-ver15`.

Now that you have learned about the SQL Server data type system, it is time to extend the conceptual model from *Chapter 3, Normalizing Data*, into a quantified physical model.

Quantifying the data model

There are a number of reasons for discussing the SQL Server data type system in the amount of detail we did. In this section, we will look closer into the importance of properly choosing data types. We will have a look at the following:

- Data quality
- Query functionality
- Performance
- Database size

We discussed data quality and query functionality in the previous section. You cannot accidentally store text in a numerical column or store a string that looks like a date but isn't a valid date in a date column. Also, storing numbers as numerical data provides you with the ability to perform computations with the data.

It should also be clear from the previous section that the data types you choose have a big impact on the query performance of your database.

In the next section, we are going to provision an Azure SQL database. To do so, we need to know (roughly) how big our database is going to be. We probably also want to know how fast the database is going to grow. We need to quantify our data model. There are two parts to quantifying the model. You need to estimate the size of the database, and you need to know the expected use of the database.

Estimating the database size

For this information, we need to know the data types of all the columns we are going to create.

Have a look at *Figure 4.2*. We could create an **Entity Relationship Diagram (ERD)** using rectangles, as shown in *Figure 4.2*:

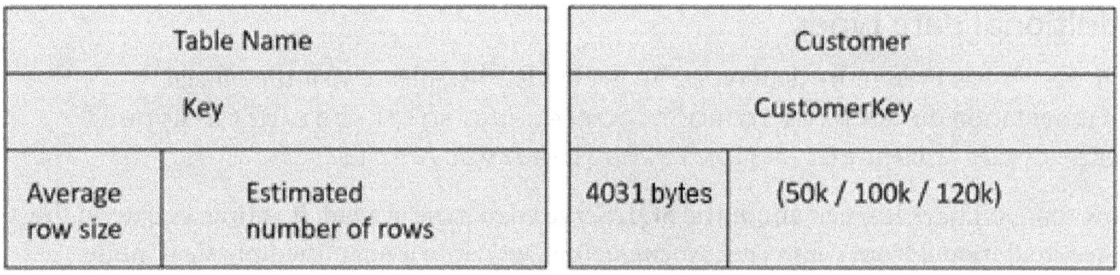

Table Name		Customer	
Key		CustomerKey	
Average row size	Estimated number of rows	4031 bytes	(50k / 100k / 120k)

Figure 4.2 – Quantifying tables

The left-hand side of *Figure 4.2* shows the concept, while the right-hand side is an example. We draw a rectangle for each table and we specify the table name and its primary key. Then we calculate the average length in bytes for a single row and document that number in the bottom-left box in the rectangle.

To estimate the average length of a row, you need to know the data type you chose for the columns and the actual values you need to store. Have a look at *Figure 4.3*:

Column	Data type	Length
CustomerID	INT	4
CompanyName	NVARCHAR(50)	37
DateFirstPurchase	DATE	3
CreditLimit	MONEY	8
...	...	...
Total		105

Figure 4.3 – Calculating the average row size

Figure 4.3 shows an example table with the first few columns. The first column, **CustomerID**, has been defined as **INT**. An **INT** data type is always 4 bytes in size, as we learned in the previous section. The second column, **CompanyName**, is slightly more difficult. The data type is **NVARCHAR(50)**, which means that the longest name we can store can be made up of 50 characters. To make an estimation of how many bytes we actually use, you need to know the average length of the customers names. If you already have the data, you can just calculate that value from your existing data, or perhaps the internet can help. Otherwise, you "guess" based on your experience.

After writing down the length of each column, you know the average row size in bytes. That brings us back to *Figure 4.2*. In the bottom-right corner of the rectangle, we estimate the total number of rows we expect to store in the table. In this case, we try to figure out how many customers we have at the moment and how many new customers we expect to get in the future.

In Azure, you do not want to make your database too big. That wastes money. One of the benefits of using the cloud is the ease with which you can let your database grow by the time you actually need it.

If you multiply the average row size by the number of rows you expect, you get the expected table size. Adding all sizes of all tables will give you a first impression of how big the database will be. However, we need to investigate expected usage patterns as well. The expected usage patterns may lead, for example, to the use of indexes in the database. Indexes cost extra storage.

Analyzing expected usage patterns

Before actually implementing the database, we need to know more about its intended use. Patterns of interest are as follows:

- Which tables are queried regularly?
- Which columns are queried regularly and which columns are regularly used together?
- Which tables are regularly joined together?
- Which columns are used to search rows by?
- Which columns are used in sorting and aggregating data?
- For heavily used tables, what is the ratio between reading from and writing to the table?

We spent a lot of time normalizing our data in *Chapter 3, Normalizing Data*. This created an ideal conceptual model. But we might need to tune the design a little for optimal performance in the real (not theoretical) world. Let's explain this using a couple of examples.

Suppose, for instance, we have a `Customer` table that we read from very often. Also suppose that the table has 50 columns, of which 10 are used very often and the other 40 almost never. We could split this table into two separate tables and call them, for example, `Customer` and `CustomerAdditionalInfo`. The `Customer` table only has the 10 columns that we use often, while the `CustomerAdditionalInfo` table gets the other 40 columns. All of a sudden, the queries that you perform a lot use a much smaller table.

Another example could be the other way around. You find that you always need to join two tables to get an additional column from the second table. Even though normalizing data says the columns should be in separate tables, you may want to add them to the same table. Instead of performing a join very often, you now query a single table.

Before you start to completely redesign your tables, see whether indexes can help. If you search a lot in the `CompanyName` column, create an index on that column. If you often use just two or three columns together, you can consider a covering index. You learn about indexing in the *Indexing* section later in this chapter.

Apart from the patterns described here, you need to look at how many concurrent users you expect and how complex the queries are that you expect. Together, they determine for a big part how much compute power your database needs. Luckily, it is easy to use cloud databases to adjust the assigned compute power. If you don't assign enough, you will get bad performance. If you assign too much, you pay too much.

> **Note**
> As a best practice for cloud resources, start small, monitor usage patterns, and increase allocated resources only when necessary.

We get back to the usage patterns in the last section of this chapter, when we discuss some SQL Server features that can help us to manage performance without paying for more allocated resources in Azure. For now, let's finally create a database!

Provisioning an Azure SQL database

It is time to get your hands dirty and create an Azure SQL database. To follow along with the examples in the rest of this chapter, you need an Azure subscription with permission to create new resources in that subscription. If you don't have a subscription, you can create a free trial subscription via `azure.microsoft.com/en-us/`:

> **Note**
>
> This section is divided into subsections. Each subsection has numbered steps starting with *step 1*. All steps from all sections must be followed in sequential order to create the database.

1. Open a browser and go to `portal.azure.com`.

2. Sign in to Azure using your Microsoft credentials.

    Logging in to Azure will bring you to your main dashboard. In the upper-left corner of the portal, you can see three vertical lines. This is the menu icon. When you hover your mouse over this icon, a popup appears saying either **Show portal menu** or **Hide portal menu**, depending on the current state of the menu. The top item of the portal menu is **Create a resource**. See *Figure 4.4*:

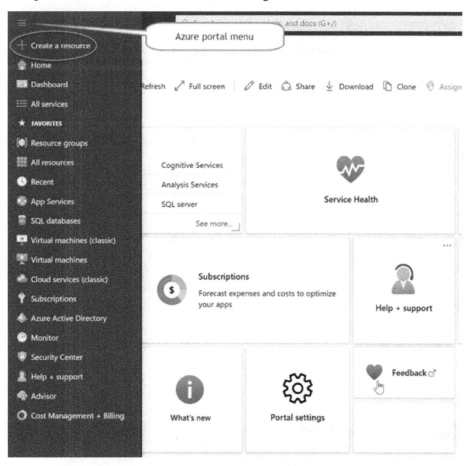

Figure 4.4 – Azure portal menu

3. Open the portal menu if necessary and click on **Create a resource**.

4. Select the textbox that currently reads **Search the Marketplace**, type SQL Database, and hit *Enter*.

5. Click, from the list of tiles that fill the screen, on the tile that reads **SQL Database**.

6. Click, on the page that opens, on the **Create** button.

   After clicking **Create**, you should see the page shown in *Figure 4.5*. Your screen might look different because Microsoft changes the layout of the Azure portal regularly. However, most options should be available:

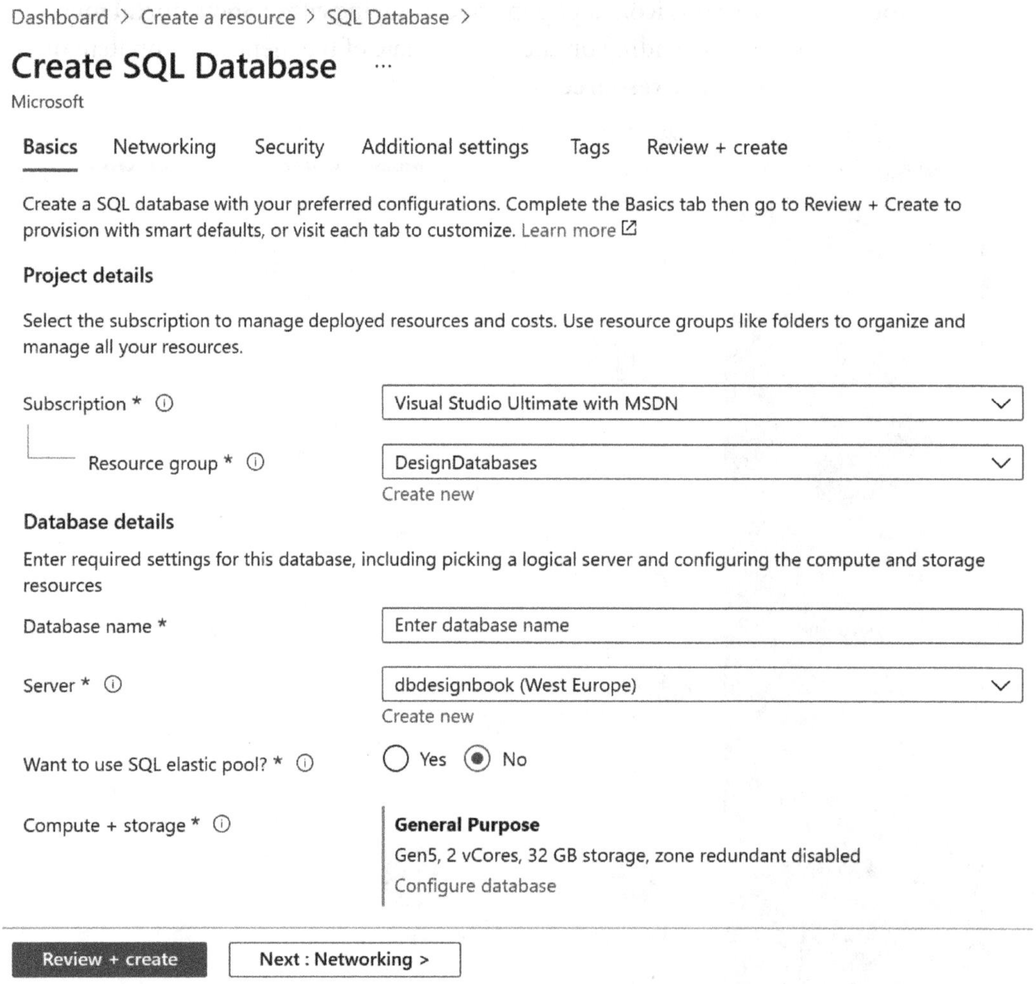

Dashboard > Create a resource > SQL Database >

Create SQL Database   ...
Microsoft

Basics   Networking   Security   Additional settings   Tags   Review + create

Create a SQL database with your preferred configurations. Complete the Basics tab then go to Review + Create to provision with smart defaults, or visit each tab to customize. Learn more ☑

Project details

Select the subscription to manage deployed resources and costs. Use resource groups like folders to organize and manage all your resources.

Subscription * ⓘ | Visual Studio Ultimate with MSDN   ⌄ |

         Resource group * ⓘ | DesignDatabases   ⌄ |
                                       Create new

Database details

Enter required settings for this database, including picking a logical server and configuring the compute and storage resources

Database name * | Enter database name |

Server * ⓘ | dbdesignbook (West Europe)   ⌄ |
                                       Create new

Want to use SQL elastic pool? * ⓘ   ◯ Yes ⦿ No

Compute + storage * ⓘ   **General Purpose**
                                     Gen5, 2 vCores, 32 GB storage, zone redundant disabled
                                     Configure database

[Review + create]   [Next : Networking >]

Figure 4.5 – Creating a SQL database; the Basics tab

If you want to create a new SQL database, there are a couple of settings you need to provide. For each resource you create in Azure, you need to specify the subscription in which you want to create the resource. This is where Microsoft will charge you for the service.

The second setting is the resource group that the new resource should be made in. A resource group is a management object. You can assign permissions to people on a subscription level, as well as on a resource group level. You may, for instance, create a resource group called `Development` and provide all your developers with permission to create new databases within this group. The resource group does not affect the functionality of the resource you create.

7. Click on the **Create new** link below the drop-down list of the resource group setting.

8. Type in `DesignDatabases` and then click **OK**.

    The next step is to provide a name for the database you are creating.

9. Name your database `Northwind`.

    The next setting is the database server. You cannot have a SQL database without a server. SQL Server has a long on-premises history where we have a SQL Server instance that is a container for one or more user databases. When applications need to connect to a database, you provide the instance name to set up the connection. Server, in this case, is another name for instance.

    In Azure, when you use an Azure SQL database, you don't have a "real" instance anymore. But Microsoft wants to keep your experience of working with SQL databases to stay as close as possible to how it was done on-premises. So, we need a server to be able to set up a connection to the database when we start working with the database. For now, we are going to create a new server.

10. Click on the **Create new** link below the drop-down box of the **Server** setting.

11. Enter `dbdesignbook`, appended with your initials, as the name of the server.

12. Choose a name of your liking for **Server admin login** and then choose your own password.

13. Choose the location closest to where you live.

    There are a couple of remarks to make about these options. First, notice that the complete server name will be whatever you enter appended with `.database.windows.net`. In my case, the SQL server I create is named `dbdesignbook.database.windows.net`. This name should be globally unique. You cannot create a server with this name as long as I keep my server in Azure. That is why you should append your initials (or something else if appending your initials still means your server name is already in use by someone else).

    Note that you cannot use uppercase letters. Also, most special characters are not allowed.

    The other interesting setting here is **Location**. It obviously defines where in Azure your database will reside. Keep your database as close to the applications using the database as possible. For starters, this will reduce the latency between the database and the applications, which will benefit overall performance. Another important argument is that you pay for data that leaves an Azure data center. So, if your application is in Asia and your database is in Europe, you pay for each query because data is transferred from Europe to Asia with each response of the database.

    You may also have to look at legal requirements for where your data may or may not be stored.

    Notice that Azure also has some requirements regarding the password you choose. Make sure to remember the name and password you choose here. By using these credentials, you can later log in to the server. Using these credentials, you will have full control over the server.

14. Keep **No** selected for the **Want to use SQL elastic pool?** option.

    SQL elastic pool is an interesting option. We will explain elastic pools after the next step.

15. Click on the **Configure database** link next to the **Compute + storage** setting.

You now get to the page shown in *Figure 4.6*:

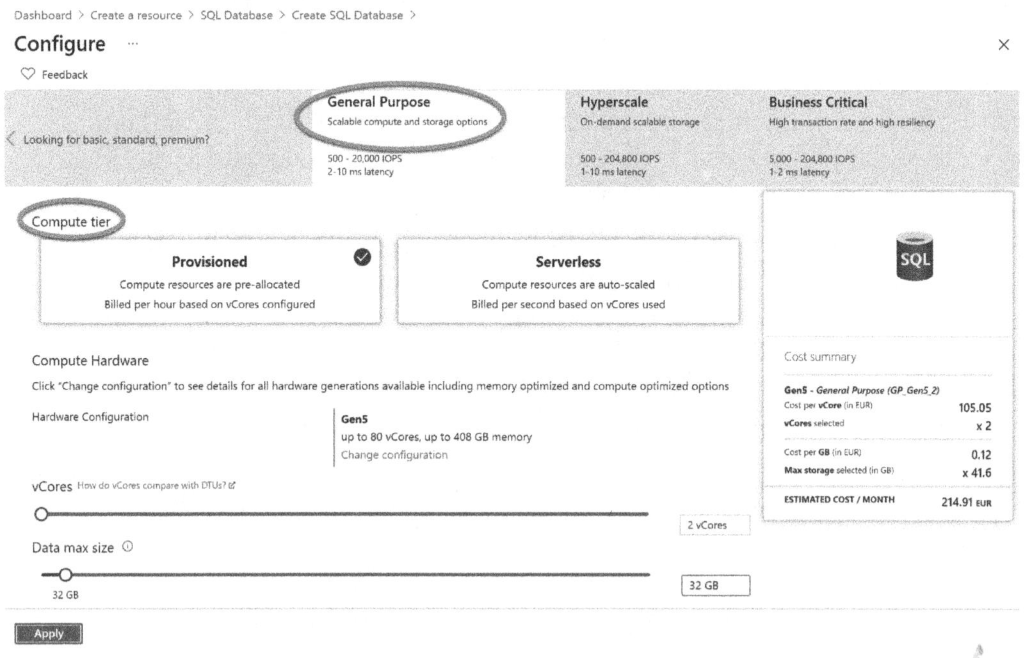

Figure 4.6 – Configuring a SQL database

There are a lot of choices to be made on this page. For now, we stay on the **General Purpose** tab. We start by discussing the compute tier that you need to configure (see *Figure 4.6*).

Provisioned versus serverless

The first choice to make is between a database using **Provisioned** resources or a **Serverless** database. When you open the **Configure** page, the **Provisioned** option is selected. This means that you will reserve dedicated resources for your database. Moving a bit down on the **Configure** page, you see two slider bars, one to configure the number of vCores you want and the second to configure the maximum size of your database.

These settings are why we had an extensive section on data types and estimated workload to start this chapter. The more concurrent users you expect, and the more complex queries you expect, the more compute power you will need. You get more compute power by provisioning more vCores. **vCores** stands for **virtual cores**, where you can think of a core as a computer's processor. With more vCores, you get more compute power, which means your database will be able to handle a bigger workload.

After selecting the number of vCores, you can select the maximum size of your database. With 2 vCores, the maximum size equals 1 TB. This can increase to 4 TB if you provision more vCores.

A database configured to be provisioned gets the storage and compute power you configure here. Those resources are dedicatedly reserved for your database. This means they are always at your disposal, even when nobody is actually using the database. But having reserved resources in Azure means you are paying.

In the screenshot of *Figure 4.6*, you can see on the right-hand side of the page that the estimated cost for the default settings is €214.91 per month. Again, you pay this amount irrespective of whether you are actually working with the database. If your database has a steady 24/7 workload, this is perfectly fine. But suppose your workload varies from hour to hour and sometimes the database might not be used at all for several hours. It would be a waste of money to pay for reserved resources that you are not using.

This is where the **Serverless** option comes in. If you click on the **Serverless** tile to select this option, you can see two slider bars for the vCores option, one for the maximum number of vCores you want to use and one for the minimum you want to use. Azure will now start monitoring your workload continuously and it will adjust your settings between the maximum and minimum settings in real time. As soon as the database starts to be used less, Azure will scale down the resources it is using. As soon as it starts using fewer resources, you start paying less money.

It gets even better with the **Enable auto-pause** option. With this option, you get to specify an inactivity period. After this period has expired with no usage of the database at all, the database will be paused automatically. You still pay for the data stored in Azure, but you stop paying for compute resources in their entirety.

Too good to be true? The **Serverless** option comes with a couple of drawbacks. To start with, Azure now needs to monitor the workload on your database and that does not come for free. When your workloads turn out to be stable over time and Azure never automatically scales your database up or down, **Serverless** is more expensive than **Provisioned**. Only choose **Serverless** with a varying workload.

Azure will only scale up when it sees resources being stretched thin. This means that you may already have observed performance-related issues. It may, for instance, then take some time for data to get loaded in memory. So, it might take a while before all problems are resolved. When you require a guarantee with regard to performance, you need to configure **Provisioned**.

Of course, you might have a scenario with a predictable but varying workload, for instance, a database that is heavily used during office hours but handles just a couple of queries during the night. You could choose a provisioned database and use Azure Automation to automatically add resources to your database just before people start coming into the office to remove those resources again at the end of the day. You need to configure this yourself, whereas with the serverless option Microsoft is doing the scaling for you. But doing it yourself does give you more control and might be cheaper. Serverless is best for irregular, unpredictable workloads.

The following table provides a short recap of the differences between serverless and provisioned:

Provisioned	Serverless
Fixed resources allocated	Allocated resources vary
Stable, predictable query performance	Varying query performance
Best/cheapest for steady, predictable workloads	Best/cheapest for irregular, unpredictable workloads

Once you have created the database, you can always adjust the number of vCores and the maximum data size. Don't over-provision a database. Over-provisioning is throwing money away. When a database is under-provisioned, you can come back and adjust the settings according to needs. There are some considerations when changing the service tier of an existing database. Refer to the official documentation for details: docs. microsoft.com/en-us/azure/azure-sql/database/single-database-scale.

Notice that in between the tiles for **Provisioned** and **Serverless** and the slider bar to configure the number of vCores, you find a section where you can select the hardware to use.

When you click on the **Change configuration** link, you can choose other hardware to use. The options that are available depend on the settings you have chosen so far. Microsoft bases the hardware they use, as well as the mix of storage, memory, and CPU, on an average **Online Transaction Processing (OLTP)** workload. In most cases, that is just fine. But you may have a workload that uses more compute that requires more CPU power than average. Or your databases may use more memory than average. In both cases, you can select a different hardware configuration here to fit your workload.

Before you spend more money and choose more expensive options for your database, ask yourself why you need those options. A badly designed database requires more resources to run the same workload than a well-designed database. That is why we have spent three and a half chapters on design before discussing all the options here. Adjusting the design is cheaper than choosing a bigger database.

The other side of that coin is that it is really expensive (and sometimes even virtually impossible) to change the database schema of an operational database. You get one chance to do it right. You can easily adjust the Azure settings once you are already operational.

Which hardware configuration you need is difficult to say. The best way to find out is to test it yourself.

Now that we know what hardware choices we can choose from, let's look at the difference between the vCore payment model and the DTU-based payment model.

vCores versus DTU

If you look back at *Figure 4.6*, you can see that the **Configure** page has four main tabs. When you open the **Configure** page, the second tab is selected. This is the **General Purpose** tab. For most databases, this is what you need. However, there are three alternatives. The first tab (the link that reads **Looking for basic, standard, premium?**) shows database options that are available under the DTU licensing model.

When SQL databases were first introduced, there was no notion of vCores. You had to provision **DTUs**, or **Database Transaction Units**. A DTU is a combination of I/O, CPU, and memory, the resources that a database uses on a server. Provisioning more DTUs means you get more resources assigned to your database. You get better performance of that database and the database will be able to handle bigger workloads. Of course, more DTUs means you pay more.

Before you use the slider bar to provision DTUs, you first choose between the **Basic**, **Standard**, and **Premium** service tiers. The **Basic** service tier is mainly intended for development, although you may, and can, run production workloads using **Basic**. The main differentiator between **Standard** and **Premium** is the number of **I/O operations per second** (**IOPS**) you get and the latency that you have during I/O. For the exact details, please visit docs.microsoft.com/en-us/azure/azure-sql/database/ service-tiers-dtu.

Databases in the DTU-based purchase model are simpler (have fewer options). When you start with Azure SQL databases, this might be an advantage. DTU-based databases also have more options for smaller databases, making them a cheaper option than vCore-based databases.

The advantages of the vCore model are as follows:

- It provides more flexibility in how you provision your database.

- It provides better insights into what you pay Microsoft for (transparency).

- It is better suited for larger databases, both in size and workload.

As a best practice, choose the vCore purchase model unless you have a good reason to do otherwise. For this book, cost is the primary reason for choosing otherwise. A **Basic** database will be enough for what we intend to do with it. **Basic** only costs a couple of euros per month.

The final setting on the **Configure** page (at the bottom of the page) is whether or not you want your database to be zone-redundant. By default, Microsoft creates three copies of each database. In the case of any hardware failure, Azure automatically switches the workload to an available copy. This mechanism provides availability to your database.

Within Azure data centers, Microsoft has created Availability Zones. Two availability zones are independent of each other. This means that when an entire zone goes down, there is still a good sense that the other zone is still functioning well. When you spread the replicas of the database over multiple availability zones, you are protected against a lot more potential issues than just hardware failures. You get more uptime for your database. Of course, you pay extra for this service.

We have already discussed a lot of settings that you need to choose wisely. From a performance standpoint, testing is the best way to reach an optimal configuration. From then on, make sure to monitor your database. The best configuration today might not be a good configuration at all in a year from now. Start small (cheap) and scale up when needed.

From a cost perspective, it is clear by now that you pay for what you get. The actual price may vary from one data center to the next, and may vary over time. I advise you to look up actual pricing information on the official Azure website once you start to make these choices.

Hyperscale and Business Critical

Apart from the **General Purpose** databases that have now been discussed and the DTU-based databases, Azure also provides the option to create a **Hyperscale** database or a **Business Critical** database.

The first thing to know about **Hyperscale** is that it uses a different architecture than the other service tiers. This means that you cannot switch from a **General Purpose** database to a **Hyperscale** database, or vice versa. As the name suggests, **Hyperscale** is for very large databases. Its main differentiators are as follows:

- You can create databases up to 100 TB in size. **Hyperscale** will autoscale to the storage required. A **General Purpose** database can grow to up to 4 TB.

- **Hyperscale** has a faster transactions log that makes transaction commit times shorter. This results in faster overall performance.

- You can quickly create read-only replicas to offload part of the workload to a replica. You can configure up to four replicas.

- You can scale up and down the number of vCores used without the impact that such an action has on **General Purpose** databases.

- Backups and restores are much faster (minutes instead of hours) than for **General Purpose** databases.

The **Business Critical** service tier (comparable to the **Premium** DTU service tier) offers two main benefits over **General Purpose** databases:

- By using local storage instead of the remote storage used by **General Purpose** databases, combined with faster storage, **Business Critical** databases have less I/O latency. This means that it can accommodate much bigger workloads because low latency increases the overall throughput.

- **Business Critical** databases are built on four-node clusters providing high availability. Also, the nodes are divided over different availability zones, increasing the availability further. The replicas can be made available to offload read workloads to the replicas.

After all these options, let's go back to creating a database.

1. Click on the link that reads **Looking for basic, standard, premium?**.

2. Select **Basic** and then click on **Apply**.

Choosing the payment model and provisioning enough resources is an important part of creating a SQL database. There is, however, an extra consideration to make when you are going to create lots of databases, such as, for instance, in a **Software-as-a-Service** (**SaaS**) scenario when you create a separate database for each customer you have. That option is the elastic pool as discussed in the next section.

Elastic pool

There is one setting left to discuss on the **Basics** tab of the **Create SQL Database** page: **Want to use SQL elastic pool?**.

Elastic pool is somewhat comparable to the serverless option that we saw earlier, but for a pool of databases instead of for a single database. As mentioned previously, the serverless option is cheaper with a varying workload, whereas the provisioned option is cheaper for a steady workload.

You may have a scenario where you have 100 databases that each individually have a varying workload. But if you look at the average workload of all databases combined, the workload turns out to be steady. When one database is heavily used, the other isn't used at all, and vice versa.

This is the ideal situation for elastic pools. With elastic pools, you opt for provisioned resources that are shared among multiple databases in the pool. Azure allocates the resources dynamically to databases that are used heavily by taking those resources away from less-used databases. By using elastic pools, you can provision far fewer resources than the combined maximum resources required per database. Because the resources are provisioned, it is cheaper than using the serverless option for each database individually. You find these types of scenarios very often in multi-tenancy **SaaS** apps.

Networking

From a database perspective, you now choose the most important settings. But there is more to do. You need to think about connectivity from applications to your database. So, the next thing to worry about is networking:

1. Click on the **Next : Networking** button to get to the **Networking** tab of the **Create SQL Database** page.

    A database, by itself, is nothing. You have applications that work with the database. Those applications have to be able to connect to the database to send queries and get result sets back. This means that we need a network that we use to communicate over.

    As you can see on the **Networking** tab, there are three options:

- **No access**
- **Public Endpoint**
- **Private Endpoint**

The **No access** option doesn't make any sense. It simply means that you will configure networking later.

2. Select the **Public Endpoint** option.

    Public Endpoint means that this database will be reachable over the internet. That does not mean that everyone can get access. There are two extra layers of security. The first is that Azure SQL Server uses IP whitelists to allow network traffic only from computers with known IP addresses. The second security layer is that you need to be able to authenticate with SQL Server. You have to be a defined user in the database in order to be able to get a connection.

3. Click on **Yes** for both options in the section on firewall rules.

    The first setting allows other services within Azure to connect. You need this option, for example, when you want to use this database in Power BI reports published to the Power BI service. The Power BI service is an example of an Azure service.

    The second setting adds your current IP address to the whitelist mentioned earlier. This allows you to connect to the database from the computer you are currently working on. Later in this chapter, we need to work with the database using tools such as **SQL Server Management Studio** (**SSMS**) or Azure Data Studio, which we install locally. Note that if you want to connect to this database at a later time from another location, or in scenarios where your workstation got assigned a new IP address, you will have to add that new IP address to the firewall rules using the database settings in order to connect to the database.

Instead of using a public endpoint, you could use a private endpoint. In Azure, we can create **virtual networks** (**vNets**). We can then create endpoints on that network. Services that have an endpoint for a network can communicate over that network. But only services connected to that network can use that network. It is outside the scope of this book to explain networking. Using private endpoints is the safest option because we have the most control over who can, and cannot, connect to the database.

After configuring the network settings, you can go ahead and create the database, or you have some additional settings that you need to consider. We will look at these additional settings in the next section.

Additional settings

Even though we will accept all the default settings on the **Additional settings** page, it is worth discussing the options you find on this page.

Click on the **Next : Additional settings** button to get to the **Additional settings** tab of the **Create SQL Database** page.

The **Additional settings** tab has three settings that you may configure. These settings are as follows:

- **Data source**
- **Database collation**
- **Maintenance window**

Let's go through them in order.

Data source

In the **Data source** setting, you need to choose between the following options:

- **None**: **None** will create an empty database.
- **Backup**: **Backup** allows you to create a new database from a backup of a database that exists, or existed, in the same Azure subscription. Microsoft automatically creates backups of all the databases you provision in Azure. The service tier of your databases defines how long a backup is kept. Restoring a database is as simple as creating a new database and choosing the **Backup** option here. You get a list with all available backups.
- **Sample**: The **Sample** option creates the `AdventureWorksLT` example database.

Database collation

The next setting is an important one that you already learned about in the section on *alphanumerical data types*. You have to define the collation for this database. To recap, the collation defines the code page (alphabet) used for columns defined with either `CHAR` or `VARCHAR`. It also defines case sensitivity and accent sensitivity, both of which have an impact on the evaluation of `SQL WHERE` clauses and the sorting order. Changing the collation afterward is not possible. You have to do this right now.

Maintenance window

The last setting provides you with the opportunity to specify the most convenient time from your perspective that Microsoft can do maintenance on your database. Microsoft needs to keep their software up to date and compliant. That needs work. Your database will be available during maintenance; however, failovers to other hardware may occur. Applications could lose connections during those failovers. Applications should always implement retry logic where connections to Azure databases are concerned. To minimize the impact of this maintenance, you get the specify a time that suits you best, for instance, during the night when your database is used the least.

By now we are almost done. There is only one more page, **Tags**, to worry about.

Tags

The next page in the entire sequence of settings is the **Tags** page. This allows you to add some metadata of your own to the database you are creating:

1. Keep all default settings on the **Additional settings** tab.

2. Click on the **Next : Tags** button to get to the **Tags** tab of the **Create SQL Database** page.

Tags have no functional meaning. They simply allow you to add descriptive metadata to the resources you create in Azure. You could, for example, create a tag with the name `Environment` and the value `Training`. Or, you could create a tag with the name `Department` and the value `Academy`. It is a way to document why you created this resource, or who should pay within your organization for this resource, or anything else that makes sense to you.

Review + create

As the name of the last page suggests, there is nothing more to do except to review your choices and actually create the database:

1. Click on the **Next : Review + create** button to get to the **Review + create** tab of the **Create SQL Database** page.

2. Click on the **Create** button to actually (finally) create the database.

While waiting on the portal to actually create the database, we need to make one final remark. There are three ways in which to create resources in Azure:

- Use the portal.
- Use scripting.
- Use Azure resource templates.

Creating a resource using the portal, as we did in this section, is a good approach when you need to learn about the resource. It is good for demos and proofs of concept. However, clicking on an interface is error-prone, and it is costly when you have to do it very often.

You always want to automate whatever you do. This is a way to document the resources you create and it is easier to repeat. You just re-execute your automation process. You can use either PowerShell or the Azure CLI to script out everything we did in this section, or you can create a template. That is basically a JSON file with all the settings we chose in it. Based on this template, you can create other databases later on. You can download the template files by clicking on **Download a template for automation** at the bottom of the page. Unfortunately, automating resource creation is outside the scope of this book.

It may take a while for the database to be provisioned. The page does show you some progress information. Once the database has been created, the page shows a button that takes you to the database. You don't have to wait for this.

Connecting to the database

Now that we have a database, it is time to work with it. There are a number of tools we can use to work with the database. To manage existing databases, the same goes as for creating new ones. You can use the Azure portal or you can use scripting. Templates only apply when you create a new resource. For developer-related tasks, such as creating the tables we want, it is easier to use specialized database tools. Let's look at a couple of options that we have at our disposal.

Azure portal

For the purposes of this book, we chose Azure Data Studio as the tool we use. But we will start by using the Azure portal:

1. Log in to the Azure portal.
2. Show the Azure portal menu by clicking on the menu button (three vertical lines) in the upper-left corner of the portal.

3. Click on **All resources** to bring up a list of all your Azure resources.

4. Click on the SQL server, **dbdesignbook**.

    You now see the SQL Server blade. Remember that we created both a SQL server and a SQL database in the previous section. The SQL Server blade is where you find non-database-specific settings, such as the server IP firewall rules, for instance.

5. Click, in the menu on the left-hand side of the SQL Server blade, on **SQL databases**.

6. Click on the **Northwind** database to open the database blade.

    You now see the database blade. You could have opened the database blade directly from the list of all your Azure resources from *step 3*. The database blade shows you database-specific settings.

7. Click, in the menu on the left-hand side of the database blade, on **Configure**.

    Notice that this is where you can scale up or scale down your database when required.

8. Click, in the menu on the left-hand side of the database blade, on **Query editor (preview)**.

9. Log in using the name and password you created in *step 12* of the *Provisioning an Azure SQL database* section.

A blade opens that allows you to enter SQL code and to run that code against your database. On the left-hand side, it shows you a list of tables, views, and stored procedures in the database. Of course, our database is still completely empty. This query editor is pretty basic. You probably want a more advanced tool to interact with your database.

Azure Data Studio

Microsoft has multiple tools to develop and manage databases. The most well-known is probably SSMS. You can download SSMS for free. SSMS is a Windows-based tool mainly designed for database management and the execution of ad hoc code.

Microsoft also has Visual Studio, which is their development tool. There is a free version that you can download and use for database development.

More recently, Microsoft introduced Azure Data Studio. This tool is platform-independent (also runs on operating systems other than Windows) and is usable for multiple Azure data services, not just for SQL databases.

We will use Azure Data Studio in this book:

1. If you don't have Azure Data Studio already, go to `https://docs.microsoft.com/en-us/sql/azure-data-studio/download-azure-data-studio?view=sql-server-ver15` to download and install it.

2. Open Azure Data Studio.

    You should now see the **Welcome** page of Azure Data Studio. As you can see in *Figure 4.7*, Azure Data Studio has a menu on the left-hand side of the screen. We have a button for connections, where you can register multiple connections to all sorts of data services. Azure Data Studio will remember connections for the next time you use the same database from Azure Data Studio. The first step to perform in Azure Data Studio is to make a connection:

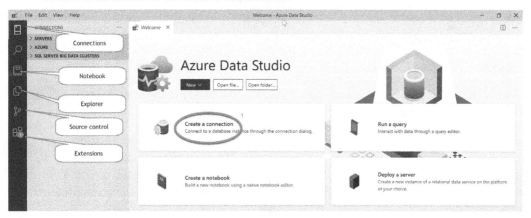

Figure 4.7 – Azure Data Studio

3. Click on **Create a connection** in the middle of the **Welcome** page (see *Figure 4.7*).

4. In the **Connection Details** pane, enter the following information:

    a. Connection type: **Microsoft SQL Server**

    b. **Server**: The name you chose in *step 11* of the *Provisioning an Azure SQL database* section (`dbdesignbook` plus your credentials) appended with `.database.windows.net` (for example, `dbdesignbook.database.windows.net`)

    c. **Authentication type: SQL Login**

    d. **Username**: The name you chose in *step 12* of the *Provisioning an Azure SQL database* section

    e. **Database: Northwind**

5.   Click on the **Connect** button. You should now see the screen of *Figure 4.8*:

Figure 4.8 – Northwind database in Azure Data Studio

As you can see in *Figure 4.8*, you now have a connection to the database. Azure Data Studio already shows a hierarchy with objects that this database contains. We also have options to create new queries and new notebooks to start interacting with the database.

We are ready to go. It is time to implement tables in our database. We will do so in the next section. But let's first finish this section with two quick remarks.

We will not make this a tool tutorial. You will get to know Azure Data Studio by using it. However, we do recommend that you familiarize yourself with the tool.

There is one remark that is worth making here. In *Figure 4.7*, you can see that there is a menu option for **Source control**. Azure Data Studio can connect to GitHub. This allows you to save all your files in GitHub in the cloud. Within GitHub, you have version control over your files and you can work together with multiple coworkers on the same projects. GitHub and source control are beyond the scope of this book, but I strongly recommend that you look into this.

Data definition language

Data definition language is the part of the SQL language that you use to manage objects in a database. By manage, we mean the following:

- CREATE
- ALTER
- DROP

You need to be able to create new objects, most notably, new tables. You also need to be able to alter existing tables or drop existing tables. The same holds for other objects, such as views and stored procedures. That is why we have three statements. Let's start with CREATE TABLE.

Creating a table

You create a new table using the CREATE TABLE statement:

1. Click, in Azure Data Studio, on the **New Query** button to open a new query file (see *Figure 4.8*).

2. Enter the following code (you can find the code in the Create table Product.sql file in the downloads):

```sql
CREATE TABLE [dbo].[Product]
(
ProductID INT IDENTITY (1,1) NOT NULL,
ProductName NVARCHAR(40) NOT NULL,
SupplierID INT NULL,
CategoryID INT NULL,
QuantityPerUnit NVARCHAR(20) NULL,
UnitPrice MONEY NULL,
UnitsInStock SMALLINT NULL,
UnitsOnOrder SMALLINT NULL,
ReorderLevel SMALLINT NULL,
Discontinued BIT NOT NULL
);
```

3. Click on the **Run** button just above the script to execute this code.

4. Open the **Tables** folder in the **Connections** pane on the left-hand side of the screen and verify that you now have a table called **dbo.Product**. You may have to refresh the folder. Click on the **Tables** folder with the right mouse button to do so.

5. Click, in the Azure Data Studio menu, on the **Explorer** button. Right-click on your query file to save it.

There are a few things to note about this CREATE TABLE statement.

Schema and table name

Directly after the CREATE TABLE keywords, you provide the name of the table to create. This name should be unique. If there is already a table with the same name, you will get an error.

The name in the example consists of two parts separated by a dot. Within SQL Server, we can create schemas. Schemas provide a way to group related tables. The schema name in which we create a table is the first part. The actual table name is the second part. In this example, we create a table called Product in a schema called dbo.

The main benefit of schemas is that they provide a security boundary. You can provide users with permissions on schemas. Those permissions are inherited by all objects in that schema. In that way, you don't have to set permissions on a table-by-table basis but you assign the requisite permission at the schema level.

You always have schemas. The dbo schema is a predefined schema that exists in every SQL database. It often serves as a default schema. You can create your own schemas using the T-SQL CREATE SCHEMA statement. The following code will create a schema called Sales:

```
CREATE SCHEMA Sales;
```

From now on, you can create tables in this new schema called Sales.

You do not always need to use a two-part name. In many cases, using just Product leads to the same result as using dbo.Product. However, using a two-part name is better and should be regarded as a best practice.

Columns

The CREATE TABLE statement is a comma-separated list of column definitions. Each line in the statement describes a single column, starting with the name of the column. The names follow the design we created during normalization.

A name cannot exceed 128 characters in total. A name cannot start with a digit; it must start with a letter. Some special characters, such as the underscore, are allowed. When you use spaces in names, or you used reserved words such as table as names, you need to enclose the name in square brackets, [].

Data type

After the column name, you specify the data type of the column. That is why we started this chapter with a section about the data types you can choose from.

Identity

In the line defining the `Product ID` column, the word `IDENTITY(1, 1)` is added after the `INT` data type. The `IDENTITY` specification tells SQL Server to automatically generate unique values for this column each time a new row is inserted into this table. The first time, SQL Server will use the value `1`. Each following row gets a value one higher than the previous row. That is what the `(1, 1)` part means. This doesn't make the `ProductID` column the primary key. But using `IDENTITY` only makes sense in combination with a primary key constraint. When combined with a primary key constraint, the identity specification makes `Product ID` a surrogate key. We will make `Product ID` a primary key later in this chapter.

NULL/NOT NULL

Each line ends with either `NULL` or `NOT NULL`. This specification determines whether or not a column should always have a value. Specifying `NULL` means that the column may be left empty. The value of the column is then said to be unknown. With the `NOT NULL` specification, each row needs to have a valid value for this column. In our example, this means that there should be a product name for each product (row), but you may leave the `SupplierID` column empty when adding a new product to the table.

Create the `Customer`, `Employee`, `Order`, `OrderDetail`, `ProductCategory`, and `Shipper` tables. You can find the definitions in the `Create all tables.sql` file in the downloads.

Once you have a table, you can always make changes to it.

Altering a table

You can add or remove columns, you can change the data type of columns, and you can add or remove constraints. Be aware that doing so on a table that already stores a lot of data might be a long-running operation. Changing an empty table is easy and will always be fast. But once you have data stored in the table, all these operations might have a big impact. Try to do it right the first time.

Now that we have the tables that we want, it is time to think about the keys. We didn't implement any keys in the tables. You can add keys to tables immediately on creation by adding the keys definition to the CREATE TABLE statement, or you can add the key later using the ALTER TABLE statement. We will add both primary keys and foreign keys to our tables. We will also add some other constraints to our tables.

Primary key

Let's start by adding primary keys to our tables. In subsequent sections, we will add foreign keys to the tables followed by other constraints:

1. Open a new query in Azure Data Studio.

2. Add the following code:

```
ALTER TABLE dbo.Product
ADD CONSTRAINT PK_Product PRIMARY KEY CLUSTERED
(ProductID);
```

3. Execute the code by clicking on the **Run** button.

    This code alters the existing dbo.Product table. It adds a constraint named PK_Product to the table. The constraint is of the primary key type. The ProductID column will be the primary key.

    Whenever you create a primary key, SQL Server will automatically create an index for you. SQL Server knows two types of indexes – clustered and nonclustered. You will learn about indexes later in this chapter. If you omit the CLUSTERED keyword, you will still get a clustered primary key. CLUSTERED is the default.

    It is a good idea to use naming conventions. Starting the name of a primary key with the PK_ prefix makes it immediately clear that you are dealing with a primary key.

    We could have added the primary key when creating the table. Let's add another table and define the primary key directly in the CREATE TABLE statement.

4. Open a new query in Azure Data Studio.

5. Add the following code and execute it (the code can be found in CREATE TABLE Supplier.sql):

```
CREATE TABLE [dbo].[Supplier]
(
        SupplierID       INT IDENTITY(1,1)    NOT NULL,
```

```
       CompanyName        NVARCHAR(40)        NOT NULL,
       ContactName        NVARCHAR(30)        NULL,
       ContactTitle       NVARCHAR(30)        NULL,
       Address            NVARCHAR(60)        NULL,
       City               NVARCHAR(15)        NULL,
       Region             NVARCHAR(15)        NULL,
       PostalCode         NVARCHAR(10)        NULL,
       Country            NVARCHAR(15)        NULL,
       Phone              NVARCHAR(24)        NULL,
       Fax                NVARCHAR(24)        NULL,
       HomePage           NVARCHAR(255)       NULL,
       CONSTRAINT PK_Supplier PRIMARY KEY CLUSTERED
   (SupplierID)
   );
```

As you can see, the syntax of adding the primary key directly in CREATE TABLE is the same as adding it later using ALTER TABLE.

6. Add primary keys to all the tables. You can find the code in the Add Primary Keys.sql file.

7. Save all your code files.

Pay special attention to the primary key of the dbo.OrderDetail table. It has a combined primary key. In between the braces, both columns participating in the key are mentioned, separated by a comma.

Foreign key

Now that the tables have primary keys, it is time to relate the tables to one another by adding foreign keys:

1. Open a new query in Azure Data Studio.

2. Add the following code and execute it (the code is available in the Add Foreign Keys.sql file):

```
ALTER TABLE dbo.[order]
ADD CONSTRAINT FK_Order_Customer FOREIGN KEY (CustomerID)
    REFERENCES dbo.Customer (CustomerID);
```

The syntax follows the same pattern as adding a primary key. You specify the table to alter, the fact that you want to add a constraint named FK_Order_Customer, and that it is a foreign key constraint. After the FOREIGN KEY keywords, you specify the column that is the foreign key. This can be a comma-separated list of columns when you have combined keys. You then end the foreign key definition by specifying the table along with the foreign key references and the primary key in that table. The column you specify doesn't have to be a primary key; however, SQL Server needs to know that it holds unique values. You ensure that it does by making it the primary key or by adding a unique constraint to the column. SQL Server raises an error when the column is not unique.

The foreign key has a couple of additional settings that you can specify. Look at the next bit of code:

```
ALTER TABLE dbo.[order]
WITH CHECK ADD CONSTRAINT FK_Order_Customer FOREIGN KEY
(CustomerID)
    REFERENCES dbo.Customer (CustomerID)
        ON DELETE CASCADE
        ON UPDATE CASCADE;
```

First, notice WITH CHECK before ADD CONSTRAINT. As with the primary key, we could have added the foreign constraint directly in the CREATE TABLE statement. If you add it later with an ALTER TABLE statement, there might already be data in the tables involved in this relationship. What should SQL Server do with that data?

Remember that we create foreign keys because of the referential integrity they provide. This means in our example that from now on, every time a new order is added to the database, SQL Server will check whether the customer ID you provide exists in the dbo.Customer table. If it does not exist, SQL Server raises an error and will not add the new order.

Adding WITH CHECK will make SQL Server check all existing data for referential integrity. It will not create the foreign key if there is data violating the referential integrity. WITH CHECK is the default behavior. You can change CHECK to NOCHECK to disable the check. SQL Server will now create a non-trusted foreign key. Referential integrity is tested for all new data added to the database.

We also added ON DELETE CASCADE and ON UPDATE CASCADE to the statement. These ensure that SQL Server will automatically delete or change all orders from a customer when that customer is deleted or changed. Instead of CASCADE, you could have specified NONE, DEFAULT, or NULL (see *Chapter 1, Introduction to Databases*). NONE is the default.

3. Add foreign keys to implement the ERD shown in *Figure 4.9* (you can find the code in the Add Foreign Keys.sql file):

4. Save all your code files:

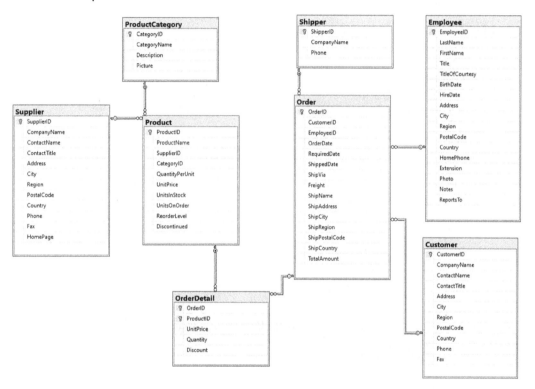

Figure 4.9 – ERD Northwind

With all keys in place, you have guaranteed the uniqueness of rows in a table and provided referential integrity as well. The next step is to see whether or not you want or need to implement other constraints. In our case, we will add a couple more constraints to the Northwind database we have created so far.

Other constraints

In *Chapter 1, Introduction to Databases*, we talked about other constraints than just the primary key and the foreign key. Let's implement an example of a default constraint, a check constraint, and a unique constraint.

To start with the last one, a unique constraint is a lot similar to a primary key. You can use unique constraints for candidate keys that did not become the primary key. In our database, we want our product names to be unique:

1. Open a new query in Azure Data Studio.

2. Add the following code and execute it (the code is available in the Add Constraints.sql file):

```
ALTER TABLE dbo.Product
ADD CONSTRAINT UX_
ProductName UNIQUE NONCLUSTERED (ProductName);
```

As for a primary key, SQL Server makes an index when creating a unique constraint. In this example, we ask for a NONCLUSTERED index.

3. Also, add the following constraints (the code is available in the Add Constraints.sql file):

```
ALTER TABLE dbo.Product
ADD CONSTRAINT CK_UnitPrice CHECK (UnitPrice > 0);
```

```
ALTER TABLE dbo.[Order]
ADD CONSTRAINT DF_
Orderdate DEFAULT SYSDATETIME() FOR OrderDate;
```

You may want to implement more constraints than the ones shown here. This section is just to show an example of each type of constraint we have in SQL Server. Feel free to add more constraints to your database.

Dropping a table

You can (of course) delete entire tables when you need to. The following statement deletes the dbo.Product table. Note that all data stored in the table is deleted as well. Also note that there is no option to reverse this action other than restoring the entire database from a backup:

```
DROP TABLE dbo.Product;
```

Inserting data

In the previous section, we have seen the CREATE, ALTER, and DROP TABLE statements. Once you have created tables, you can start adding rows to the table. The best way to do so depends on the scenario. The basic SQL statements that you use to manipulate data are the following:

- INSERT
- UPDATE
- DELETE

Together, these statements are called **Data Manipulation Language** (**DML**). We will use the regular INSERT statement to add some data to our database:

1. Open the dbo.Customer.Table.sql file from the downloads and execute it. The first two lines of this file are shown here:

```
SET IDENTITY_INSERT [dbo].[Customer] ON
```

```
INSERT [dbo].
[Customer] ([CustomerID], [CompanyName], [ContactName],
ContactTitle], [Address], [City], [Region], [PostalCode],
[Country], [Phone], [Fax]) VALUES (1, N'Alfreds
Futterkiste', N'Maria Anders', N'Sales Representative',
N'Obere Str. 57', N'Berlin', NULL, N'12209', N'Germany',
N'030-0074321', N'030-0076545')
```

Remember that we created a surrogate key. The CustomerID column is created with the IDENTITY specification, meaning SQL Server will provide values for this column. However, in our current scenario, we want to maintain the customer IDs that we used in the system from where the customer data is exported. SET IDENTITY_INSERT overrides the default identity behavior and lets us use our own customer IDs. Notice that the script you executed ends by setting IDENTITY_INSERT off again.

The second line of code is a basic INSERT statement that inserts one row of new data. It specifies the table in which a new row has to be added. It lists all the columns that we will provide a value for. It then lists the values we want to enter.

The INSERT statement is meant for entering individual rows or a couple of rows at best. For entering multiple rows, the BULK_INSERT statement is better optimized. We will take a look at the BULK_INSERT statement in *Chapter 10, Designing and Implementing a Data Lake Using Azure Storage*.

2. You can find a script in the downloads comparable to `dbo.Customer.Table.sql`, which you can use to add some data to the other tables. Execute them all. Make sure to first add data to tables that are referenced by foreign keys before adding data to tables that hold the foreign key.

   This book is not intended to be a book about the SQL query language. We will not discuss all the other DML options.

3. Execute the following SELECT statement to test the database:

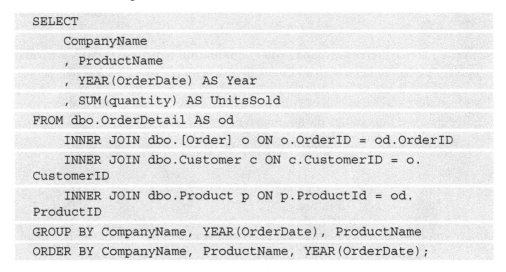

```
SELECT
    CompanyName
    , ProductName
    , YEAR(OrderDate) AS Year
    , SUM(quantity) AS UnitsSold
FROM dbo.OrderDetail AS od
    INNER JOIN dbo.[Order] o ON o.OrderID = od.OrderID
    INNER JOIN dbo.Customer c ON c.CustomerID = o.CustomerID
    INNER JOIN dbo.Product p ON p.ProductId = od.ProductID
GROUP BY CompanyName, YEAR(OrderDate), ProductName
ORDER BY CompanyName, ProductName, YEAR(OrderDate);
```

Executing this SQL SELECT statement should return rows that you added when you successfully executed the scripts mentioned in *step 2*.

Now that we have created a database and successfully inserted some data, it is time to have a quick look at database performance.

Indexing

Congratulations! You have provisioned and implemented an Azure SQL database. There are a lot of extra features that you need to learn in relation to SQL Server. A good book to start with is *SQL Server 2017 Developer's Guide*, by Dejan Sarka, Miloš Radivojević, and William Durkin. There is one topic that is too important to not devote some time to in this regard: indexing.

Roughly speaking, indexes make reading data from a database faster, but slow down writing to databases. So, you have to be careful with the indexes you create. But without indexes, you are almost sure to experience performance problems.

There are two basic kinds of indexes:

- Clustered indexes
- Nonclustered indexes

Clustered index

A clustered index could be described as an index-organized table. When you create a clustered index, you store the actual rows sorted on the index key in a structure called a B-Tree. The index *is* the table. A clustered index is not a structure or object that you create next to the table you need to index. You change the physical storage structure of the table into a B-Tree structure. From the set-based theory discussed in *Chapter 1, Introduction to Databases*, you know that tables do not store rows in any specific order. That is no longer true with a clustered index. After we created clustered primary keys in the *Altering a table* section, SQL Server started storing the rows sorted on the primary keys. This speeds up applications that search for rows using the primary key, a very common scenario.

Because the clustered index is actually the table itself stored in a specific way, there is just one clustered index per table.

You can compare a clustered index to an old-fashioned printed telephone directory. It is optimized to retrieve a phone number based on someone's last name and residence. But what if you have a phone number and want to know who it belongs to? Even though the answer is in the directory, you probably don't even bother to start looking. You know it is going to be a long-running query. A telephone directory heavily optimizes one query, but is a really bad choice for other types of queries. A clustered index optimizes queries that search on the index key really well. It doesn't help with other queries.

Nonclustered index

Because you never have just one type of query on a table, you can add extra indexes to it. These are nonclustered indexes. **Nonclustered indexes** are extra storage structures next to the actual table.

In the same way that you can compare a clustered index with a telephone directory, you can compare a nonclustered index with an index in a textbook. At the end of the book, you will find some additional pages. They don't store any new information. There is an ordered list of keywords printed on those extra pages. Each keyword goes together with a number. This is the page number of the page on which you can read more about the keyword.

When, for instance, you create a nonclustered index on the `ProductName` column, SQL Server will create an ordered list of all product names and store the location of where to find the row describing that product. When a customer searches for a product by its name, SQL Server can now use this nonclustered index to quickly retrieve the row that has been searched for.

Looking back at the database we created in this chapter, you can see that all tables have a clustered index on the primary key. The `Product` table has an additional nonclustered index on the `ProductName` column because you created a unique constraint on the column.

Defining the proper indexes is a really important task. It requires a lot of technical knowledge regarding how SQL Server works internally and how applications use the database. Luckily, we get some help from SQL Server.

Automatic tuning

Azure SQL Database has an automatic tuning feature that can create indexes for you. You can find this feature on the **Northwind** blade in the Azure portal. You can see the page in the following screenshot:

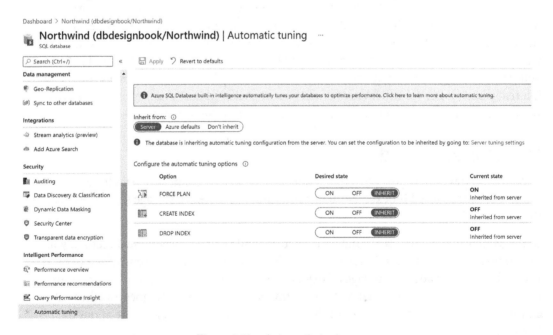

Figure 4.10 – Automatic tuning

Having learned a little about indexes, that concludes this chapter. As we said at the start of this section, there is a lot more to know when implementing a database. Unfortunately, that is outside the scope of this book.

Summary

We started this chapter by explaining the most common data types in SQL Server. Choosing the right data types is an important step in the design process. Data types determine both the efficiency and functionality of the database.

We then provisioned a database in Azure SQL Database. Choosing the right performance tier is important. If you under-provision a database, you will experience bad performance. If you over-provision a database, you are wasting money. Because the way a database is used over time may change, you need to monitor the database and rescale whenever necessary. It is important to note that a well-designed database requires fewer resources to handle the same workload as a badly designed database.

After provisioning a database, you need to create tables. This is where you implement the design you made.

Now that you have learned a lot about SQL databases, it is time to look at designing a NoSQL database in the next chapter.

5
Designing a NoSQL Database

Relational databases have been the dominant database type for decades. But in the last 10 years, there have been ever more use cases and scenarios where relational databases can't deliver what we need anymore. They lack the scalability, performance, and functionality that some modern applications require. That is why, in this chapter, we will look at alternatives to relational (SQL) databases.

We will discuss the following topics in this chapter:

- Understanding big data
- Understanding big data clusters and partitioning
- Azure Cosmos DB
- Key-value databases

Understanding big data

What is big data?

Well, here's one practical definition:

> *You have big data when you need to innovate to do with your data what you need to do.*

This means that if your biggest database is an Excel sheet where you have reached the limit of 1 million rows, you have big data. In this case, you need to use another application, possibly on other hardware, to keep working with that data.

This definition is more of a joke than a real definition. However, there is a kernel of truth in there, as well as a warning. When you want to do with your data what companies have been doing with their data for years, you are probably better off using a relational database. Some companies implement big data solutions because they feel they have to. They worry that if they don't, the competition will. But without a solid business case for big data, you are better off using a well-known solution on proven technology. The big data world is moving very rapidly. There are new platforms, new architectures, and new best practices almost every year. That makes big data implementation difficult and expensive. But you do have a solid business case when a relational database cannot deliver what you need.

A better description of what big data is would involve the four Vs. Depending on who you ask, big data is defined by three, four, five, or even more Vs. For this book, I will use four Vs to describe big data. Those four Vs are as follows:

- Volume
- Velocity
- Variety
- Variability

Big data is when one or more of the characteristics volume, velocity, variety, and variability pushes relational databases to (or over) their limits, leading to scalability, performance, and functionality issues.

Volume

Volume seems to be the characteristic for which *big* data is actually named. The name *big data* suggests that we are dealing with lots and lots of data. The question is, what do we mean by lots of data? We saw in *Chapter 4, Provision and Implement an Azure SQL Database*, that an Azure SQL database of the Hyperscale service tier can scale up to 100 TB.

There is no limit to how many Hyperscale databases you can make in Azure. In *Chapter 8, Provision and Implement an Azure Synapse SQL Pool*, we will discuss Azure Synapse, which is a relational database platform that handles even larger amounts of data. Just having a lot of data is very likely not the sole reason that you have big data. At the same time, we should consider both the storing and the processing of data. The amount of processing power needed when data volumes grow might warrant the use of big data technologies.

Having data and using data might be two different things altogether. If you need to keep data that you are not using regularly, storing that data inside a relational database might not be the most cost-effective solution. Azure Blob storage will be far cheaper than database alternatives. The combination of the amount of data and what you intend to do with it is what makes big data what it is.

Velocity

By **velocity**, we mean the speed with which new data comes into existence. This speed has increased tremendously in recent years. More data than ever is being "born" digitally, with ever more devices producing more and more data. When relational databases emerged, data entry typists would manually enter data into the database. As people were rather slow at this, that did not put a lot of strain on databases. Now, with millions of sensors all around the world generating data, it's an entirely different story.

Manually entered data has changed a lot as well. We used to have some people employed to do data entry. Today, customers on your website enter data by creating accounts and placing orders. For a popular website or app, we could be talking about thousands of people entering data at the same time. People are still slow, but there are a lot more of them.

There are lots of scenarios where relational databases struggle to keep up with the insert rate that we demand from them. Big data systems might be better scalable to keep up with the insert rate. But even these systems may reach their limits, in which case, we need to rely on other solutions such as queuing or streaming data using, for instance, Azure Event Hubs and Azure Stream Analytics.

Variety

Another thing that has changed since the introduction of relational databases is the type of data we work with. A relational database is well suited to working with numbers, strings, and dates. You only have to look back at the previous chapter to understand this. Financial transactions are characterized by a transaction amount, the day the transaction took place, and data such as the names of customers, products, or ledger entries. Using these data types, we can process and store payments, analyze those payments, and create reports for things such as sales by month. But today, there is more **variety** in the data we work with.

A JPEG file is digital data. When, for instance, you have a JPEG picture showing some people, you might want to write a query asking how many people are in the picture and their average age. This is "just a query," and you have the data to answer the question. This, of course, requires some machine learning, and the question comes with some uncertainty, but it is still just a query on some data.

Pictures are an example of unstructured data. Other examples could include things from music and movies to PDF documents. A table in a relational database is an example of structured data. We know exactly what we have because we know which columns are in the table and we know the data types of the columns. The data and metadata are combined.

Semi-structured data is, as the name suggests, somewhere in between structured and unstructured. Examples include CSV files and XML files. A CSV file is very often structured, in the sense that we know how many values are stored and what data types they are. The main difference is that the metadata resides in the head of the person working with the data. The data and metadata are separate. With XML, we add the metadata to the data. However, an XML schema is flexible in that it allows different elements and attributes to be used by different rows. The metadata isn't fixed but instead changes all the time.

Relational databases are all about structured data, but they might struggle with semi-structured and unstructured data. Relational databases are well suited for transactional data, such as orders and payments, even when that leads to a lot of data. Storing pictures in a relational database, however, is not really efficient. More importantly, using T-SQL in your database will get you nowhere when trying to do meaningful things with pictures. You end up paying for a relational engine that stores your data (cost) inefficiently and does not provide you with any functionality.

Variability

In all the previous chapters, we worked with tables and the columns that the tables were made of. We call that schema-on-write, as explained in *Chapter 1, Introduction to Databases*. By schema-on-write, we mean that we create our tables first, and only after we have created the tables are we able to store data. A problem arises if you do not know beforehand what data to store. This means that you don't know which columns to create, and you can't create a table if you don't know its columns.

> **Note**
> Schema-on-write means you create tables first before you can store data.

The Northwind database that we created in *Chapter 4, Provision and Implement an Azure SQL Database*, was easy. Northwind sells delicacies from all over the world. They *only* sell delicacies, which makes all the products the same. That is, all products can be described using the same characteristics. In other words, they can be described using the same columns.

But think about a product table for a webshop such as Amazon. Back in the days when Amazon was still only a book store, managing such a table would have been easy. Each book is described by its ISBN, title, author, language, genre, and so on. But today, Amazon sells computers as well. None of those characteristics describes a computer. Amazon also sells baby food. This makes the problem bigger again.

Suppose you created a column for each possible characteristic for each possible product your webshop sold. How many columns would that be? The answer: probably more than your database can handle. Besides, for all products, most of the columns would not be used. That would be very inefficient because the database would be reserving space for those columns regardless.

The problem described here is **variability**. Each product needs different columns, and every time you start selling different products, your table has to be adjusted again. We need a kind of flexibility that a table with fixed columns can never provide.

Another example of variability is where data may change over time. Consider a scenario where you buy thermometers to monitor the temperatures in rooms in office buildings. In each room, you have a thermometer that sends time, temperature, and location data to you every 5 minutes. You create a table with three columns to store these measurements.

After a while, some thermometers break, so you buy new ones. Only, the version you have is not for sale anymore, so you buy the new version. However, this one sends location information as GPS coordinates, whereas the old one just sent a string with a manually entered room name. Later down the line, you might get thermometers that also send you humidity information, meaning you now have four columns of information. Eventually, you will have all sorts of different devices sending you different information in all sorts of different forms. Having a predefined table with fixed columns will not give you the freedom to accommodate these differences. You need to be more flexible in your solution. There is too much variability.

Extend the scenario to a multinational with a lot of buildings all over the world. You would now start to have a lot of data coming in at high speed. Suppose you need to react to the measurements in real time to keep all rooms at a certain temperature with a minimal amount of energy being wasted. This adds velocity and volume to the mix. NoSQL solutions can help.

Big data, then, involves the following factors:

- We have more types of data than we used to have.
- Data is coming at us at much higher rates than it used to.
- We need to perform complex analyses of the data we have.

These points mean that we need to treat our data differently. This is not new. An OLTP workload, for instance, asks for a different design of our database than an OLAP workload. When either the specs of the data itself or the usage of the data changes, we need to design our data solution differently.

Understanding big data clusters

A really important part of working with modern data solutions is the scalability of the solution. Scalability determines how well a system will keep functioning when we experience growth. Growth can mean any or all of the following:

- The system needs to handle more concurrent users.
- The volume of the data we need to store increases.
- The compute power needed increases because the query complexity increases.

The last two points are about being able to utilize more hardware resources. The main resources we need to consider are compute and storage. Compute refers to the number of CPU cores being used. Storage can mean storing data on actual hard drives or storing data in memory. In the end, data must always be stored on hard drives.

Hardware scalability is about adding more hardware resources to our database. The second part of scalability is to do with whether or not our database will actually benefit from extra hardware. This is where good database design comes into the picture again.

Hardware scalability can be done in two different ways:

- Scaling up
- Scaling out

Scaling up

Scaling up means using a bigger server. When you start with a small database, a small server might be enough to run that database. When you translate that to Azure, we are concerned with the service tier. In *Chapter 4, Provision and Implement an Azure SQL Database*, you created a DTU-based **Basic** service tier database. This is the smallest and cheapest database you can provision in Azure. When the workload grows because more concurrent users are executing more queries per second, you can increase the number of DTUs you provisioned by changing the service tier from **Basic** to **Standard**. When you need faster storage, you can change to the **Premium** service tier. You keep increasing the power of the hardware (server) you use.

Part of the benefit of using the cloud is what we call **elasticity**. Elasticity is how easy it is to scale up when needed but also to **scale down** when possible. **Scaling down** means assigning fewer resources to your database. The optimum elasticity option is using the serverless option for your SQL database. Scale up and scale down are sometimes referred to as vertical scaling.

The problem here is that there is always a limit to scaling up. At some point, your database will be running in the highest service tier that Azure has to offer. What if that is not enough anymore? You reached the limit of scale up! Apart from physical limits, you will often see that multiple smaller servers are cheaper than one big server that has the same power as the smaller servers together.

Scaling out

Scaling out means you are not using a bigger server but instead are using more servers and dividing the workload over those multiple servers. You may start with a single server. After a time, you might add a second server. Later, you might add another two servers, meaning you now have four servers running your workload. In theory, you can always add extra servers. There is always a limit to how powerful a single server can be. There is no limit to how many servers you can use (as long as your credit card cooperates).

> **Note**
> Scaling up means using a bigger server, whereas scaling out means using more servers.

Scaling out is implemented using clusters. A **cluster** is a group of servers working together as one (see *Figure 5.1*):

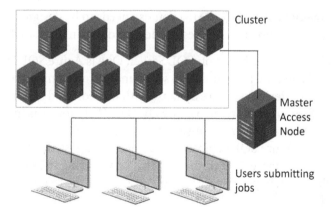

Figure 5.1 – Cluster

Each server participating in a cluster is called a **cluster node**, or just a node. Applications (users) always connect to one of the nodes. This node is called the **head node** or master node. In this way, users do not have to be aware that they are working with a cluster, nor do they need to know how big the cluster is. The head node divides the workload over the **worker nodes**. The worker nodes do the actual work. The head node's main task is to then send the final result back to the client. You can seamlessly add worker nodes to the cluster because that doesn't change anything for the clients working with the cluster. The client communicates with a single server, the head node. The head node distributes the work over the worker nodes and communicates back to the client.

Each worker node stores part of the data when we implement databases on cluster technology. Each worker node can also participate in executing queries. This means that a cluster provides scale-out for both compute and storage resources. When you remove cluster nodes from the cluster, that's **scale-in**. Scale-out and scale-in are sometimes referred to as horizontal scaling.

Adding nodes to a cluster adds risk to the system. With more servers, there is a bigger chance of one of the servers experiencing hardware failures. Adding scalability does not make sense when it means that the system is down all the time. All cluster technology should have built-in resiliency. This means that it can detect failures on individual nodes and recover from them automatically and without downtime.

To successfully work with clusters, you need to have two things:

- Cluster-aware software

- A good partitioning strategy

Cluster-aware software

The **database management system (DBMS)** you use needs to be designed to operate on a cluster. Most traditional relational databases are designed to run on a single server.

Azure has two database services based on SQL Server. One is Azure SQL Database. This service is a traditional database implementing scale-up by provisioning more DTUs or more vCores. The other offering is Azure Synapse with its dedicated SQL pool. Synapse dedicated SQL pool, formerly known as Azure SQL Data Warehouse, is in all senses a relational database. You can create a Synapse SQL pool in a SQL server, as you always need a server to host an Azure SQL database. It can even be the same SQL server you use for normal Azure SQL Database. You connect to this server using the same tools as with a normal SQL database. Then, you use the same T-SQL language to interact with Synapse dedicated SQL pool.

The difference is that under the hood, 60 databases are created, called distributions, for each SQL pool. The data you store in Synapse dedicated SQL pool is automatically divided over those 60 distributions. Each distribution is hosted on a different server when you provision the biggest SQL pool available. Or, in other words, a Synapse SQL pool runs on a cluster.

One of the characteristics of NoSQL databases is that they run on clusters. That does mean that we face an extra issue in the design phase. Suppose you store all your data for the current year on the last cluster node. Suppose then that 99% of the queries executed ask questions regarding the current year. You are now virtually back to using a single server. We need to make sure we store the data cleverly over the different nodes available *and* that we use the compute of all the nodes equally.

We should consider situations where we write a join between the **Order** table and the **Customer** table to retrieve sales by customer. How can we efficiently join the tables together when matching rows from both tables are stored on different cluster nodes? Matching rows will have to be moved to the same node before that node can perform the join operation. Data movement, however, is always a time-consuming operation. We want to avoid data movement as much as possible.

When we randomly distribute the data over the cluster nodes for storage, we do achieve scalability in terms of storage. However, we lose a lot of efficiency when it comes to actually using the data. Even though we are using cluster technology, we still do not have the scalability we are after.

Partitioning

Partitioning is the way in which we distribute our data properly over the cluster nodes. When defining a partitioning strategy, we take both storage and compute into account. Only when we implement a well-chosen partitioning strategy will we get the scalability that big data clusters promise. You need to avoid all new data that you need to insert into a database being stored on the same cluster node. Likewise, you don't want all the data you read to be stored on the same cluster node. In both cases, you create a bottleneck on that cluster node. You pay for the cluster but you are not taking advantage of it. The exact implementation of how you define partitioning might vary slightly from system to system. Let's first look at Synapse. We will explain how Cosmos DB tackles partitioning in the *Cosmos DB partitioning* section.

When we look at Azure Synapse dedicated SQL pool again, we can distinguish three ways to partition data. You can partition data on a per-table basis. We'll look at how to create tables in Synapse in *Chapter 8*, *Provision and Implement an Azure Synapse SQL Pool*. Synapse dedicated SQL pool uses the term distribution instead of partition to avoid confusion with partitioned tables. Partitioned tables are a SQL feature that we can implement both in Azure SQL Database as well as in Synapse dedicated SQL pools. The three ways to partition data in Synapse are as follows:

- Hash distribution
- Replicating data
- Round robin

Hash distribution

Using **hash distribution**, rows are distributed over the system based on a hash value calculated from a distribution (partition) key. You need to choose the column that will be the hash key. For instance, in a **Customer** table, you could choose the **City** column to be the hash key. From each value for **City**, a hash function calculates that city's hash value. The hash value is a number in the range from 1 to the number of cluster nodes (it is actually slightly more complex but this simplification makes it easy to understand what is happening inside Synapse). For Synapse, this means a value in the range of 1 to 60, because Synapse has 60 distributions. The value "Amsterdam" might, for instance, be hashed into the number 5, whereas the value "London" might be hashed into the number 10. Each time a customer from Amsterdam is stored, the row will be added to the fifth distribution. Each time a customer from London is stored, the row will be added to the tenth distribution. All customers living in the same city are stored in the same distribution.

With a hash partitioning strategy, you can get the most out of your big data cluster. The inverse is true as well: when you choose the wrong column to be used as a hash key, it will hurt the scalability of your database tremendously. Suppose in the previously mentioned example of hashing by city that 90% of your customers come from Amsterdam. This now means that 90% of the rows are stored in a single distribution and the other 10% of the rows are distributed over 59 distributions. This is really inefficient. You need to choose a key with a lot of distinct values as the hash key in order to efficiently distribute your data over the cluster.

Imagine that you choose the **CustomerID** column as the hash key instead of the **City** column. Now suppose you also choose the **CustomerID** column as the hash key for the **Order** table. Because Synapse always uses the same hash functions, rows with the same **CustomerID** value will be stored in the same distribution regardless of the table.

This means that when you join the **Customer** table with the **Order** table, matching rows in **CustomerID** are guaranteed to be in the same distribution. Each cluster node can do the join on its part of the data without any rows having to move first. All partial joins together form the end result of the join. Both the storage and the compute of your cluster are utilized efficiently.

Figure 5.2 shows a schematic of how hash-distributed tables work in Synapse. This figure is from the official Microsoft documentation:

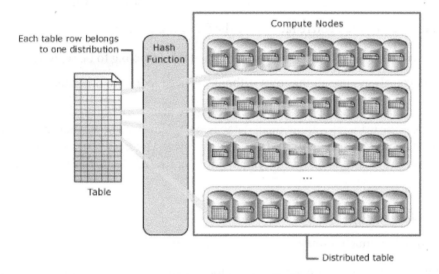

Figure 5.2 – Hash-distributed table

As said, in Synapse there are three ways to distribute data over the cluster. Next to hashing, you can choose replicating data as an alternative.

Replicating data

Replicating data is an alternative to hashing. A replicated table stores all its rows on all the distributions. In the case of Synapse dedicated SQL pool, this means that you create 60 identical copies of a table and store a copy in each distribution. You should choose this option for small tables only.

Replicating data is ideal for a date dimension in a star schema database. You will learn about date dimensions and star schemas in *Chapter 7, Dimensional Modeling*. Suffice to say now that a date dimension is always small and almost always used a lot in queries. By replicating the entire table to all nodes, a join with another table can always be executed efficiently using all nodes regardless of the partitioning strategy you choose for the other table. In the previous example, we could only join the **Customer** and **Product** tables efficiently when they had the same hash key.

Round robin

Round robin will randomly distribute rows over the available distributions. This comes with the guarantee that all distributions will get an equal portion of the rows. You can get some skew (where one distribution has a lot more rows than other distributions) with a badly chosen hash key. Round robin will never lead to skew.

However, when you need to join a table that you distributed using round robin to another table that is distributed using either hashing or round robin, matching rows will (most likely) not be in the same distribution. This leads to data movement, which will slow down your queries.

Choose round robin when you're faced with the following circumstances:

- There is no usable column for a hash and the table is too big to be replicated.
- When you always use all the rows without the need to join to other tables.

The theory explained here applies specifically to Azure Synapse dedicated SQL pool. However, the same principles apply to all databases implemented on cluster technology. To design a database for scalability and cost-effectiveness, you need a strategy to distribute your data over the cluster. A badly chosen partitioning strategy leads to the following:

- Bad performance
- Lack of scalability
- Spending too much money on bigger service tiers

Now that we understand what big data stands for and what clusters are, it is time to look into Microsoft's big data database in Azure, Cosmos DB

Getting to know Cosmos DB

There are many different NoSQL databases. They all have their own special characteristics. Microsoft offers Azure Cosmos DB as their flagship NoSQL cloud database. Cosmos DB allows us to use different APIs to connect to Cosmos DB. This is to facilitate migration from different on-premises NoSQL databases to Azure. Before we go into other types of databases, we will discuss Cosmos DB's native type. Cosmos DB is a **document database**.

MongoDB is the best-known open source document database. One of the APIs that Cosmos DB supports is the MongoDB API. This means that applications can write code to Cosmos DB as if it were a MongoDB. However, be aware that Cosmos DB only mimics the API from MongoDB. It actually has a different logical structure underneath that you need to be aware of.

Cosmos DB also has a SQL API. This means that we can write `SELECT` statements to query Cosmos DB databases.

Document databases store data as **JSON** documents. Document databases do not store documents in general such as PDF documents, Word documents, and so on. That would be the field of document management systems. Document databases store data like a relational database. That means we store data about entities. The difference is that entities do not translate into tables. They might translate into JSON documents. However, more often than not, there is no one-to-one translation from entities to JSON documents. Let's have a short look at what JSON is before we start designing JSON documents.

JSON

JSON stands for **JavaScript Object Notation**. It was invented as a means of exchanging data between applications on the internet. A prerequisite was that all browsers and platforms should be able to work with JSON. That is why JSON is just plain text. You could create JSON documents using Notepad.

As the name suggests, JSON works with objects. A lot of programming languages are object-oriented, which means that they work with objects. Objects in this sense are closely related to entities. However, an object might be far more complex than an entity. Objects might consist of multiple layers of nested objects.

Relational databases often suffer from what is called an **impedance mismatch**. The application working with the database needs to join multiple tables together to get the data it needs. In *Chapter 3*, *Normalizing Data*, a really simple report led to an ERD with four different tables. That makes sense from a redundancy perspective. But every time a form is saved or a report is created, the data has to be split over the tables or joined to make a whole again. We store the data differently than how we use the data. That is what we call the impedance mismatch. All the work this involves hinders the performance and scalability of the system.

Maybe we should store the data in the same way as we use the data. And maybe we should use simple technology without a lot of overhead. Storing data as text in JSON format is just that.

JSON formats text into objects that consist of key-value pairs. A key is a string enclosed in double-quotes. A key in JSON is what would be a column in a table in a relational database. The value is the actual data you need to store. A JSON object is denoted by curly braces. You can see an example in the following code:

```
{"City":"Sydney"
, "Country":"Australia"}
```

City and Country are the keys, and Sydney and Australia are the corresponding values. Together they form an object. Key-value pairs are comma-separated.

The values in JSON have data types, as columns do in relational databases. There is a difference, though: JSON is just text, which means that everything is stored as text. Numerical data is not stored more efficiently than alphanumerical data because both are stored as text. Conceptually, we can still make the distinction. The data types are as follows:

- String
- Number (type integer)
- Number (type float)
- JSON object
- Array
- Boolean
- Null

The interesting data type from a data modeling perspective is the JSON object. Using objects, we can nest objects within each other. For instance, we can create an Order object, and nested inside Order we can have a Customer object. That would lead to this JSON:

```
{
"OrderID":1
, "OrderDate":"15-3-2020"
, "Customer":{"Name":"John" , "City":"New York"}
}
```

Notice that the value 1 for OrderID is not enclosed in double-quotes. That is not necessary because it is of a numerical data type. Also, note that the date is stored in the European format, with the day first. There is no such thing as a "Date" data type that enforces real data types and checks the values to see whether they are valid for that data type. Every check that the database does is good for data quality but bad for performance and scalability. Storing data as JSON is (partly) about making performance and scalability more important than data quality.

Another interesting data type is the array. When a key possibly has multiple values, you can store those values in a comma-separated way within square brackets. You can see an example in the following code, which describes a Book object:

```
{
"ISBN":" 9781786465344"
, "Title":"SQL Server 2016 Developer's Guide"
, "Authors":[ "Dejan Sarka", "Miloš Radivojević", "William Durkin"]
}
```

This JSON describes a single book that has been authored by three authors. You can also combine arrays and objects by creating an array of objects. When, for instance, you need to keep track of the countries each of these authors lives in, you could create an `Author` object with two keys: `Name` and `Country`. You then create a list inside which you store three comma-separated objects.

What we are showing here will seem horrible if you are used to normalizing. In essence, we are talking about the first normal form here. Instead of saying we have two entities with a many-to-many relationship, we talk about nesting related entities within each other and creating a list for repeating values. Saying that we can do this, however, is not saying that we always should. The question becomes, how should we model data for a document database using JSON?

A complete description of JSON is outside the scope of this book. You can find a good JSON tutorial at `www.w3schools.com/js/js_json_intro.asp`.

Modeling JSON

The first important observation to make is that everything is allowed as long as you follow the basic rules just described. Let's, for example, look more closely into storing information about books and authors. A book can be authored by multiple authors and an author can have authored multiple books. You can see the ERD belonging to this scenario in *Figure 5.3*. *Figure 5.3* shows the two entities involved. Because we concluded we are dealing with a many-to-many relationship, we have already created the bridge table. When you want to print all author-related information for a book, you need to join three tables. When you enter a new book written by a new author, you need to enter new rows in three different tables:

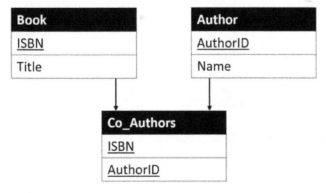

Figure 5.3 – Books and authors ERD

You can create a separate JSON document for each entity (table), as seen in *Figure 5.3*. You can also create a single document that has all the information in it. *All the information* could mean all the information about a specific book. It could also mean all the information about a specific author. With the latter option, you create a JSON document for each author you know. Within that document, you create an array of `Book` objects. You could also embed objects for `Author` inside the `Book` object. You can see the JSON for both options in the code listed here:

- Book with `Authors.json`:

```
{
    "ISBN": "9781786465344",
    "Title": "SQL Server 2016 Developer's Guide",
    "Authors": [
      {
        "AuthorID": 1,
        "Name": "Dejan Sarka"
      },
      {
        "AuthorID": 2,
        "Name": "Miloš Radivojević"
      },
      {
        "AuthorID": 3,
        "Name": "William Durkin"
      }
    ]
}
```

- Author with `Books.json`:

```
{
    "AuthorID":4,
    "Name":"Peter ter Braake",
    "Books":[
      {
        "ISBN":"9789024429554",
        "Title":"Database modelleren"
      },
```

```
{
    "ISBN":"9789024419258",
    "Title":"Leerboek Business Intelligence"
},
{
    "ISBN":"9789024404223",
    "Title":"Handboek Power BI"
}
]
}
```

Both options are valid and need to be considered.

Suppose that in 90% of your queries, you try to locate all you know about a book based on the book title. With the first listing, this query is a single GET request to the database. This means that you just made 90% of your queries simple, fast, and cheap.

On LinkedIn, however, the second JSON would probably work better. People use LinkedIn to find a person and to find out as much as possible about that person. In this scenario, you look for the author by name to retrieve all the books they wrote. The second JSON would provide everything you are looking for with a single GET request. In both cases, the JSON optimizes a specific usage.

Notice that using the wrong JSON would cost you dearly. Suppose most of your queries search on the name of the author to retrieve a list of books. And suppose further that you implemented the first version of the JSON, where one document describes a book with a list of authors embedded. This would mean that the database needs to scan through all documents to find the books you are after. Even though indexes can help to make this less troublesome, you can still imagine that using a single GET request to retrieve a single document is always preferable to scanning all documents, especially when big data is involved or the scalability of a global website is of concern.

Using embedding versus referencing

Instead of creating a single document, we can create separate documents for separate entities as well. The above examples use embedding. Either the authors are embedded in the book document or the book information is embedded in the author document. An alternative to this approach is to create separate documents for books and authors. In an author document, you add a list of just ISBN numbers. These ISBN numbers can act as foreign keys. You can use an ISBN number to find the book document holding all the detailed information of the book with the corresponding ISBN number. In this case, we talk about referencing: an attribute within a JSON document references another JSON document.

Neither embedding or referencing is a priori better. It all depends on how you use the information stored in the documents. Some arguments to use embedded objects (to nest entities within entities), as we did in the previous two examples, are as follows:

- They should be used when the relationships between entities are of a nature where one of the entities can be seen as a part of a bigger entity. You can think, for instance, of orderliness being part of an order.

- One-to-many relationships can be better described as being one-to-few relationships. For instance, books can have more than one author but never thousands of authors.

- Embedded objects hardly ever change. Think of how often the title and ISBN of a book will change over time.

- The data of embedded objects is also needed in most cases where the main object is retrieved.

The first argument will, in a lot of cases, have some overlap with the last. In both cases, the underlying idea is that one GET of a document that now retrieves all the information that's needed is more efficient than using two or more GET requests because the data needed is divided over multiple documents.

The second of the preceding arguments brings some nuance to the argument of trying to use a single GET request. If, in the second JSON listing, an author has written hundreds of books, and there is a design where we store dozens of attributes per book instead of just the title and the ISBN number, the document will become large and inefficient.

While you provision DTUs or vCores for Azure SQL databases, you provision **Request Units (RUs)** for Cosmos DB. You will learn how to do this in *Chapter 6, Provision and Implement an Azure Cosmos DB Instance*. An RU is roughly one GET request of a document with a size of 1 KB. This means that you need to provision more RUs for your Cosmos DB database when your documents increase in size. More RUs, of course, means that you spend more money. When you don't use half of the information you retrieve but you do pay for retrieving it, you are wasting money. The other side is when you oftentimes need to retrieve a second referenced document because the retrieved document references other documents instead of having more information embedded. You spend extra RUs when you need to retrieve multiple documents. It all boils down to: how do you use the information stored in the database?

Think back to our example of an author and their books. Suppose we are talking about Packt, the publisher of this book. How often do you think potential customers will be interested in a book that was published, say, 10 years ago? In our fast-moving IT world, a book that old would be good if you were a collector or interested in starting a bonfire. You could create a JSON document where you listed only the five most recent books of an author. This keeps the JSON small and efficient and will in most cases hold all the data you need. You can keep a separate list of older books that you only need to retrieve once in a while.

The third argument, that embedded objects should hardly ever change, is worth mentioning here as well. Suppose we choose to create JSON documents for books with author information embedded in them. Suppose also that we keep track of the royalties an author gets. Now, finally, suppose that a book is sold by an author who has written 10 books. Their royalties are stored in 10 different documents. This means you have to find each individual document and adjust the royalties in each of them. Again, this is really inefficient.

When you look at both JSON documents that we have suggested so far (Book with `Authors.json` and Author with `Books.json`), you can see that they both contain de-normalized data. De-normalized data always means redundant data. There is a reason for normalizing data. Performing a lot of writes in the database was one of them. De-normalizing data often has a positive effect on the read performance of a database. De-normalizing data has a negative effect on the write performance of a database. This is no different when using Cosmos DB instead of Azure SQL Database.

There are no fixed rules, like the steps you can follow to normalize data, when it comes to designing JSON documents. For the most part, the same underlying ideas hold true. You need to optimize your database for its intended workload. You need to know the entities involved and know how they are used together. Also, you need to know which data changes often and reduce data redundancy for that data. Knowing your data and how the data is used is even more important for big data databases than for relational databases.

There are more alternatives to designing JSON when it comes to our authors and books. We can refer to data instead of embedding data. Let's look into that.

Referring to objects

Instead of creating one JSON document that stores all that we know about an author or a book, we can create separate documents for each entity. You would get the following JSON:

- Author document:

```
{
    "AuthorID":4,
    "Name":"Peter ter Braake",
    "Books":["ISBN":"9789024429554",
            "ISBN":"9789024419258",
            "ISBN":"9789024404223"
            ]
}
```

- Book document:

```
{
    "ISBN":"9789024429554",
    "Title":"Database modelleren"
},
{
    "ISBN":"9789024419258",
    "Title":"Leerboek Business Intelligence"
},
{
    "ISBN":"9789024404223",
    "Title":"Handboek Power BI"
}
```

The disadvantage here is that retrieving all books for an author now requires multiple GET requests. That means you need to provision more RUs for your Cosmos DB database to be able to handle the same workload. With less redundancy in the data, writing will become more efficient and require fewer RUs. It is up to you to balance both against each other.

Notice that the ISBN number is a logical foreign key in this last example. You store all the ISBN numbers of an author's books as an array in the Author document. This ISBN number can be used to retrieve Book documents.

The JSON shown in the listing is (again) only an example of what is possible. We could have created a Book document with a list of AuthorID instances inside and separate documents for Author. Or, we could combine both solutions to have a list of ISBN numbers inside the Author document and have a list of AuthorID instances within the Book document.

The last option to mention is creating a document for every entity shown in the ERD of *Figure 5.3*. However, this will always lead to multiple round trips to the database, no matter the scenario. Creating JSON documents as though they were tables is almost never a good idea.

Embedding objects and referring to objects can be combined as well.

It is important to realize that data modeling is just as important (and sometimes even more) for NoSQL databases as it is for relational databases. The biggest difference is that we looked into designing a relational database, and only at the very end (in *Chapter 4, Provision and Implement an Azure SQL Database*) did we mention that you need to analyze how the data is used. That may lead to some adjustments in the design.

When designing NoSQL databases, the (intended) usage leads to how you set up your database, not some theoretical rules such as normalizing data. Cosmos DB will not perform well, nor will it scale well, if you define your document structure badly.

Another really important fact to note is that Cosmos DB is agnostic of the JSON content. That means that Cosmos DB doesn't know what is inside the JSON. You could add a key-value pair for `ScreenResolution` in the JSON for one product and add `Absorption` in the JSON for another product. This provides the flexibility we need when variability becomes an issue.

Cosmos DB partitioning

Cosmos DB is a NoSQL database implementation that is built to scale. That means that it uses cluster technology under the hood. This means that you need to design a partition strategy, as we discussed in the previous section.

There is no such notion as cluster nodes. Cosmos DB is a **Database as a Service** (**DBaaS**), where the underlying hardware is abstracted away. Cosmos DB distributes all data over what are referred to as partitions: logical partitions and physical partitions. Have a look at *Figure 5.4*:

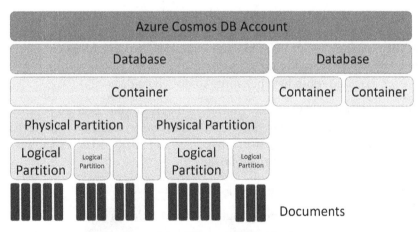

Figure 5.4 – Cosmos DB partitioning

In *Figure 5.4*, you can see vertical bars at the bottom. The yellow bars represent stored JSON documents. Each document is stored in a logical partition and each logical partition is stored on a physical partition. When you provision a new collection in a Cosmos DB database, you provision a number of RUs per second for that collection. As we have already said, the more RUs per second you provision, the more you pay, and the more resources Microsoft allocates for you. To be more exact, you get one physical partition per 10,000 RU/s that you provision.

A physical partition comes with two hard limitations:

- A physical partition can store a maximum of 10 GB of data.
- A physical partition can facilitate a maximum of 10,000 RU/s.

A physical partition stores data from one or multiple logical partitions. When the data in all logical partitions exceeds the 10 GB limit, a new physical partition is created automatically and logical partitions will be spread over the original and the new partition. You do not get more RUs automatically because you have to pay for those. An important limitation to understand is that a logical partition can never span multiple physical partitions. That means that the maximum capacity of a logical partition is also 10 GB.

Now that we know how Cosmos DB is set up, we need to set that knowledge to work and make sure our data is spread over the partitions in an efficient way.

Choosing a partition key and how to choose the correct one

It is up to you to distribute your data evenly over the logical partitions. You do that by choosing a partition key. You choose a partition key when you create a new collection in Cosmos DB. You *cannot* change the partition key afterward.

A partition key in Cosmos DB works very similarly to the hash distribution we discussed in the section about partitioning data in a cluster. You can, for instance, choose a **City** column (the **City** key inside your JSON document) to be the partition key. All customers from the same city will be stored in the same logical partition. Cosmos DB has no alternatives, such as round robin or replication, to distribute data.

Choosing the right partition key is really important. Suppose you choose a **Sex** column to be the partition key. Assuming that we have only males and females, we now have just two partitions, which means a maximum of 20 GB storage and 20,000 RUs per second. A well-chosen partition key should result in the following:

- An even distribution of data by partition size
- An even distribution of RU throughput for read workloads
- An even distribution of RU throughput for write workloads
- Enough cardinality in your partitions to mean that over time, you will not hit the physical limits of physical partitions

This means you need to choose a partition key with a lot of different values. That will lead to a lot of logical partitions that can easily be spread over multiple physical partitions. You also want to use a key that is frequently used in where clauses. It is easy for Cosmos DB to locate a document when it knows in which partition to look. That happens when the partition key is (part of) the where clause of a SQL SELECT statement used to query the database.

Using `OrderDate` as a partition key is a bad practice. All new orders would be written to the same partition. Your write workload would become skewed. You would have a so-called hot partition that all the writes go to and a bunch of partitions that do nothing. Adding more RUs to the system would not help. The one physical partition your logical partition is stored on would still have its maximum of 10,000 RU/s.

Another thing to take into account when choosing the partition key is whether you need multi-document transactions. A transaction is a single unit of work that either fails completely or is successfully executed. You don't want a new order in the system without order lines. This can happen when order lines are separate documents that fail to be written because the partition is full. You can only use transactions in Cosmos DB within a logical partition. When an **Order** document has **OrderID** as the partition key and **OrderLine** documents use **OrderID** as well, you can add an **Order** document and an **OrderLine** document as a single transaction. When, for instance, you choose **ProductID** to be the partition key of **OrderLine** documents, adding **Order** and the accompanying **OrderLine** will be independent tasks that can fail or succeed independently of one another.

As we already mentioned, the partition key is chosen when you make a new container in Cosmos DB. It cannot be altered afterward. It has a huge impact on the scalability of your system. Make sure you choose the partition key well.

Putting it together

Let's see if we can make sense from this using our previous book example where we store data about books and authors. Since the intended usage is paramount to a well-chosen design, let's make some assumptions about the data and what we will do with this data:

- We are a publisher of technical books where the time span that a book is relevant and actually sold is short. This results in the fact that for each author, only a small number of books are actively sold at a specific moment in time.

- We have thousands of books in our catalog and authors have an average of 5 books that they have written.

- When a new book of a new author is added to the database, we need to ensure the information is added as an atomic transaction.

- Each time a book is sold, we need to increment a counter that keeps track of the number of times a book is sold. We also need to increment a value that keeps track of the royalties an author should receive. The application performing this operation knows both the author and the book involved in this update.

- The main uses of the application are the following:

  a) To keep track of who we need to pay royalties to and how much

  b) To keep track of the last time royalties were paid

  c) To keep track of which authors write popular books

With this scenario in mind, the first thing to do is choose the JSON document structures to use. Since we have to update the number of books sold constantly as well as the royalties to be paid, we choose separate documents for books and authors. Author documents will have a list of ISBN numbers that reference the books that are still actively sold. Book documents hold a reference to the author. Using this setup, we keep both author and book documents small.

The next step is to think about the partitioning strategy. Since a new book from a new author should be added as an atomic transaction, we need both documents to use the same partitioning. This limits the choices that we have severely.

Since we have thousands of titles and an average of 5 books per author, both the number of distinct authors and distinct books is large. That means both are potential candidates to use as a partition key. The application knows both the author and the book whenever it needs to update the database. This also means both are still good candidates to use as partition keys as they can be used in query predicates. The main read workload seems centered around finding author information. We expect queries where information is retrieved based on the author.

This leads to the author being the best partition key to use. We explicitly choose the author name because it is known by the application and used in queries. Columns in WHERE clauses that are used as partition keys optimize the queries heavily because Cosmos DB knows where to look for the requested document. Using the author name will also distribute our data evenly over the available partitions because author names are random. It will also distribute our workload randomly and evenly over the partitions. And author names have enough distinct values to make sure we don't exceed physical limitations such as the 10 GB storage and 10,000 RU/s of physical partitions.

Note that this is a simplified example and that different assumptions or some real-world complications might lead to another column being a better partition key.

Key-value databases

With Azure Table storage (or just Azure tables), Microsoft also offers a **key-value database**. The Table API of Cosmos DB lets you write code against Cosmos DB as if it were a key-value database. This should facilitate the migration of Azure tables implementations to Cosmos DB. Azure tables have less functionality than Cosmos DB but are also cheaper. It is worth looking into key-value databases.

A key-value database stores data in values. Values can easily be retrieved using a key. It is comparable to a table with just two columns, a key column and a value column. The key is likely informative data by itself. The value is likely compound information.

Suppose you create an account on a website based on a key-value database. Your email address is used as the username. Your email address would be the key of the underlying key-value database. Using a hashing algorithm, the key determines the cluster node to store the data on. The value stores all the information you need to provide when creating your account. It can contain anything, from simple data such as addresses and names to entire pictures or JSON structures.

Examples of key-value-based applications include Windows Explorer and Apple Finder. Key-value-based applications do not necessarily always run on a cluster, but they might. Key-value databases are the most straightforward NoSQL databases around.

A workload suited for key-value databases has four characteristics:

- The database schema is straightforward, for instance, a login name can be used as the key with the account details as the value or a filename as the key with the file content as the value.

- You have high throughput. The "velocity" V of big data plays a role in your design considerations here.

- The data itself does not change regularly.

- Scalability is an important aspect of your design. A website with users from all around the world should still perform well with millions of subscribers.

The value of a key-value database is not some predefined structure. It has the same flexibility as the JSON in the previous section. It goes even further: the value doesn't even have to be JSON. It can be for one key, but it may be an image for another key. This brings the "variability" V of big data into the mix.

Now that we know what a key-value database is and when to use one, let's look into the data modeling part of setting up a key-value database.

Modeling key-value databases

The way an application uses its data is the primary concern when designing a key-value database. The intended queries are the starting point for our design. Data is always retrieved based on a key. A key-value database does not have its own query language. Using the key, the application retrieves the entire value. The application needs to understand the value and get the necessary data out of that value. Designing a key-value database, then, boils down to designing smart keys. It's almost like setting up an intuitive folder structure with an intuitive naming convention to easily find and retrieve files that are stored on your hard drive.

The key

We started by posing that an email address could be used as a key in a key-value database. Although that is not entirely untrue, we do need smarter keys than that. Look, for example, at the ERD in *Figure 5.5*. What would be the best structure to store this data in a key-value database?

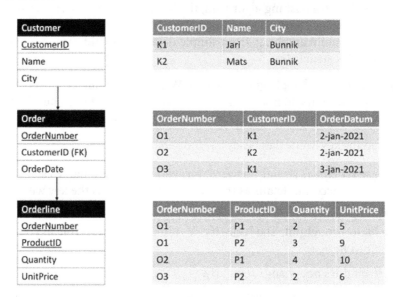

Figure 5.5 – Order ERD

The first step in designing a key-value database is creating a namespace. A namespace is comparable to a table in a relational database. Use the key to find the value within the namespace. Looking back at *Figure 5.5*, this leads to the following:

```
Customer[K1] = "Jari"
```

This code means that the K1 key in the Customer namespace has a value of Jari. With a City column in the Customer table, this would translate to the following:

```
CustomerName[K1] = "Jari"
CustomerCity[K1] = "Bunnik"
```

Using this approach, you would quickly get a lot of namespaces. Plus, an application would need a lot of round trips to the database to get all the data. With 10 columns in a relational table translated this way, you would need 10 namespaces. To retrieve all customer information, you would need to look up a value in 10 different namespaces using the same key.

You could create a JSON document to store all columns and use the JSON as the values instead of creating values for each individual column. The same arguments apply that we discussed in the section about Cosmos DB. For instance, you could create this JSON:

```
{
    "OrderNumber": "O1",
    "OrderDate": "2-jan-2021",
    "Orderlines": [
        {
            "ProductID": "P1",
            "Quantity": 2,
            "UnitPrice": 5
        },
        {
            "ProductID": "P2",
            "Quantity": 3,
            "UnitPrice": 9
        }
    ],
    "Customer": {
        "Name": "Jari",
        "City": "Bunnik"
    }
}
```

As we argued in the section about Cosmos DB, this JSON is a good choice when the application always needs all of this information and it does not change regularly. One read to the database will provide all the information needed. Using this JSON as a value would be inefficient if the application only needed part of the data. Also, if the main query were something like "Get all orders from 2020," this setup would again not be sufficient.

You don't want too many namespaces. At the same time, you don't want to store everything in a single value. Besides that, what key should we use to retrieve the JSON from the code listing? This is why there is a convention on how to create a key from data and metadata combined. Looking back at the customer Jari from our example, you could create the following key:

```
WebShop[Customer:K1:Name]  =  "Jari"
WebShop[Customer:K1:City]  =  "Bunnik"
```

There is now a generic namespace, `WebShop`. The key comprises what we would call the table, the key value, and the column name in a relational world. The table is `Customer`, the value of the key is the `CustomerID` value, which is `K1` in this example, and the column is `Name`. Together, that makes `Customer:K1:Name`. The value stored in that column is `Jari`.

Instead of using a surrogate key such as `K1`, you would be better off using a value that has a logical meaning, a value that you would use as a person to use in a search, such as the username. The preceding code could be changed to this:

```
WebShop [Customer:Jari@gmail.com:Name]  =  "Jari"
WebShop [Customer:Jari@gmail.com:City]  =  "Bunnik"
```

The queries you expect should be the starting point of the choices you make. Suppose you need to retrieve orders based on a range of dates, for example, all orders from January 2021. The key in the code fragment shown here would be a good choice:

```
WebShop [Order:20210102:Jari@gmail.com:Orderlines]  =  …
WebShop  [Order:20210102:Mats@gmail.com: Orderlines]  =  …
WebShop  [Order:20210103:Jari@gmail.com: Orderlines]  =  …
```

Notice that we use both the order date and the customer's email address to create a combined key. The value here could be some JSON with data from all order lines belonging to the key.

Be careful with a key consisting of just the date. You will probably get a hotspot in your cluster. In the preceding example, the email address will make sure the data gets distributed evenly over the cluster.

The value

The value in a key-value database can be anything. In the preceding examples, we used atomic values as the value but we also used JSON. The value could just as well be a comma-separated list of values or simply a **Binary Large Object** (**BLOB**).

Smaller values will be easier to process and are easy to cache locally to further improve performance. That makes smaller values better than (too) large values. On the other hand, you need to make sure you limit the number of times an application has to go back to the database to retrieve more information. Data that is used together should be stored together to limit the number of round trips to the database. In real life, there is sometimes a balancing act between these two best practices.

The keys are the key to success in a key-value database. The keys determine how easily you can retrieve your data *and* how the data gets distributed over the cluster. In terms of usage, key-value databases and document databases are alike. Key-value databases are often more basic, which means more of the necessary functionality should be programmed in the application.

Other NoSQL databases

Azure Cosmos DB allows us to choose between five different APIs when programming against Cosmos DB. All APIs can be used using multiple programming languages. The five APIs are the following:

- SQL
- MongoDB
- Gremlin
- Cassandra
- Table

The preferred API to use is the SQL API. This allows a programmer to use SQL queries to query the database. This leverages the power of the SQL language *and* the SQL experience that a lot of developers have. We will see some examples of using SQL in Cosmos DB in *Chapter 6, Provision and Implement an Azure Cosmos DB Instance.*

The other four APIs all have the purpose of making migrations from on-premises NoSQL implementations to Azure Cosmos DB easier. We already mentioned the MongoDB API and the Table API. Use the Table API when you have existing code that works against a MongoDB document database. Use the Table API when you are migrating an Azure tables database to Cosmos DB.

Gremlin

The Gremlin API should facilitate migrations from Gremlin databases. Gremlin is a graph database. Graph databases are specialized databases concerned mainly with querying relationships. In relational databases, many-to-many relationships are split using bridge tables because there is information to be stored about the relationship. Graph databases recognize that there is always information in relationships. Look, for instance, at LinkedIn. A big part of LinkedIn is who knows who, who posts what, and who likes which posts. As another example, have a look at *Figure 5.6*:

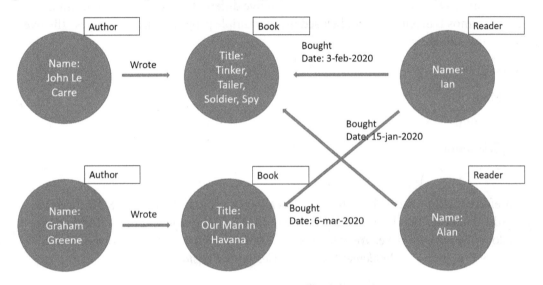

Figure 5.6 – Graph databases

Figure 5.6 was taken from www.7wdata.be and shows an example of a graph database. Each blue circle represents a row in an entity. We can recognize three different entities: **Author**, **Book**, and **Reader**. We call these nodes in a graph database. Each arrow also represents a row in the database. We call these edges. Edges have directionality, defining a relationship from a node to another node. This might be bidirectional, for instance, when two people are friends. Storing edges with directionality facilitates queries such as "Which readers read which books?" and "What else do readers who read books from John Le Carre read?". With large datasets, these types of queries will become time-consuming when done in relational databases.

A complete explanation of graph databases is outside the scope of this book.

Cassandra

The Cassandra API should facilitate migrations from Cassandra databases. Cassandra is a so-called wide columnstore database. Columnstore databases are optimized for OLAP workloads. OLAP workloads are the main topic of part 2 of this book. We will look at columnstore indexes in Azure Synapse dedicated SQL pool in *Chapter 5, Provision and Implement an Azure Synapse SQL Pool.*

Extra considerations

Now that we know how to design a Cosmos DB database (whether it's a document database or a key-value database), let's take some time to consider whether to use Azure SQL Database or Azure Cosmos DB. We didn't start this chapter with a discussion of big data and the four Vs for nothing! The JSON storage of Cosmos DB gives a lot of flexibility that comes in very handy when dealing with variability. The cluster technology and the fact that you can easily create copies of a Cosmos DB database around the globe provide enormous scalability possibilities.

At the same time, it should be noted that Azure SQL Database can scale up to many terabytes, can handle XML and JSON data, has node and edge tables to do graph database queries, and can create read-only replicas around the globe. Azure SQL Database is known and proven technology with more functionality inside the database itself. SQL databases are also very much capable of optimizing queries by taking advantage of all the metadata known about the database at compile time.

Before you choose whether to implement an Azure SQL database or a Cosmos DB database, there are two more considerations to add to the mix: polyglot persistence and concurrency. Let's start with polyglot persistence.

Polyglot persistence

In many scenarios, you will find arguments both for and against the use of Azure SQL Database and Cosmos DB. When it comes to storing BLOBs (such as pictures, movies, and music), you should also consider using Azure Blob storage (discussed in *Chapter 10, Design and Implement a Data Lake Using Azure Storage*).

A large webshop might, for instance, need Cosmos DB for the variability of the products sold and may also need the global distribution feature to store data close to where the user is. At the same time, a major part of the application will handle the orders, payments, and shipping of products. This data is very much structured in nature and can greatly benefit from the use of a relational database. You might make instruction videos available to customers on how to assemble the products once they have received them. Storing videos is best done using Azure Storage.

An application such as the one described here doesn't necessarily have to use a single type of database. Why not store the products in Cosmos DB, the payments in Azure SQL Database, and the videos in Azure Storage? Dividing data across different types of databases is called polyglot persistence.

> **Note**
> Polyglot persistence means that part of the data is stored in one type of database and other parts of the data are stored in another type of database.

There is no one-size-fits-all solution in the current data landscape. Choose the best database solution based on the specs of your data and the intended usage of the data. Using a single database is easier and probably cheaper to maintain as well. But when using a single database doesn't deliver the performance and scalability that you need, use multiple different databases alongside each other.

Concurrency

Another consideration to think about when choosing the right database is how different databases handle concurrency differently. Concurrency means that different users run queries in the database at the same time. With Cosmos DB potentially having multiple write replicas in different parts of the world, this can become rather complex.

Azure SQL databases use the so-called ACID properties to keep data consistent at all times. ACID stands for this:

- **Atomicity**: A write to the database is a single unit of work that succeeds or fails in its entirety. When you need multiple steps in code to program a change that is logically one operation, everything is rolled back when a single step fails. A developer defines logical operations by implementing transactions. When looking back, you can always say that a transaction succeeded or wasn't executed at all. There is no such thing as half a transaction.

- **Consistency**: All data is consistent after a transaction completes. For instance, you cannot have an order in the database for a non-existing customer. That would be inconsistent.

- **Isolation**: Concurrent transactions are separated (isolated) from each other, allowing each transaction to do its work independently of the other transaction. The database protects us from two transactions changing the same data at the same time and in doing so overwriting each other's data. This would lead to inconsistencies in the data.

- **Durability**: After the successful completion of a transaction, the data is durably stored in the database until the next transaction changes the data again at some later time.

There is a lot of overhead in keeping the ACID properties. Rows or entire tables must be locked when someone is working with those rows. Azure SQL Database places the "old" version of the row in `tempdb` for others to use until a transaction is really committed to the database. When two transactions try to change the same data simultaneously, one gets an error. In short, the ACID properties are great because they guarantee consistency, but they hurt the scalability because of the extra overhead incurred.

Cosmos DB uses what is called BASE instead of ACID. BASE stands for this:

- **Basic availability**: The data should be available as much as possible. When data is stored on a single server, the data will not be available when that server has a failure. Restoring the loss of a corrupt database might take hours, during which time the database will not be available. Using cluster technology, one server can go down without the entire database being unavailable. Cosmos DB databases are also easily replicated. This means that when an entire Azure region goes down, your Cosmos DB database can be available from another region.

- **Soft state**: The state a database is in can change without users writing in the database at that specific time. When Cosmos DB is replicated over multiple regions, one region can have a different version of a row than another region. This will be the case until all replicas are synchronized with each other. The version you see when querying the database depends on the region you are connected with.

- **Eventual consistency**: Cosmos DB guarantees that after you stop modifying data, all replicas of the database will eventually synchronize, after which time they will be in the same consistent state. Because Cosmos DB does not guarantee that one replica is in sync with all other replicas, it can increase write performance.

Suppose, for instance, that you sell peanut butter for €2.00 using a webshop with a global audience. You store the product information in a Cosmos DB database. To increase the responsiveness of the website, you create replicas in Asia, Europe, and America in order to have a copy of the database in close proximity to the potential buyers decreasing the latency. At 12:00 CET, you change the price of the peanut butter to €2.50 from your Amsterdam office. You are connected to the closest replica, which is the replica hosted in Europe.

Someone in China could log in to your webshop just after 12:00 CTE and still see the old price of €2.00. It might take some time to replicate the new value to the Asia data center. After refreshing their browser, the new price would be shown. But suppose your buyer refreshes the browser again, but there is an issue in the Asia data center. Because this is detected, they get re-routed to the US data center automatically. It could be the case that America is online but not synchronized with Europe (yet). Your customers now see the old price again.

This scenario is extreme but not impossible when using Cosmos DB. Availability and performance are considered to be more important than possible consistency issues such as those described here.

ACID is not better than BASE, nor is BASE better than ACID. Databases using ACID make it easier for application developers to keep their data consistent. BASE makes it easier to get good read and write performance and maintain a high level of availability at the same time.

Cosmos DB can do atomic transactions. A key thing to realize, however, is that it can only do transactions from the server-side JavaScript code using documents that use the same partitioning. This makes transactions a part of your design. The partitioning and the way you store data values in JSON documents define whether or not you can use transactions as you can in Azure SQL Database with ACID.

Cosmos DB does provide you with settings that provide control over how extreme your situation might become. In the preceding scenario, you have out-of-order reads. This means that the customer from China saw the old price even after they had seen the new price. Using consistency levels can prevent this from happening. You will learn more about consistency levels in *Chapter 6, Provision and Implement an Azure Cosmos DB Instance*.

Summary

Designing a NoSQL database follows less strict rules than designing a relational database. This follows logically from the fact that we have structured data in a relational database. We use that structure in designing the database.

NoSQL databases such as Cosmos DB should allow for more flexibility and scalability. We achieve this by letting go of strict rules and optimizing the data for its usage. The way you use the data should be the primary factor in deciding what you store together and what you store in separate documents.

Also, Cosmos DB is implemented on cluster technology. This means you have to create a partitioning strategy. You do this based on the specs of your data and, again, the way you expect to search for data.

In this chapter, you learned what big data is and when to use NoSQL databases. You learned different ways of distributing data over cluster nodes. You also learned how to choose a proper distribution strategy.

Now that you know how to design a Cosmos DB database, it is time to provision and implement one. You will do just that in the next chapter.

Exercise

You are tasked with designing the database for a new website where users can post articles. Users can do the following:

- Upload their own articles in predefined categories such as sports or science
- Add comments to articles

The application should be able to do the following:

- Show articles per category sorted by upload date
- Show all comments on a post
- Show a list of articles uploaded by a specific user

You can see the ERD in *Figure 5.7*:

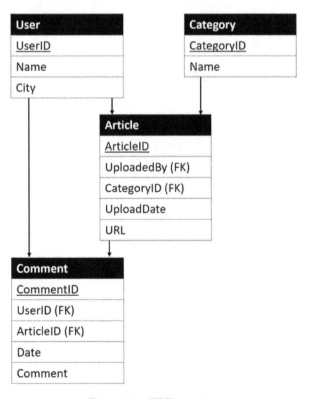

Figure 5.7 – ERD exercise

Translate the ERD of *Figure 5.7* into one or more JSON documents and explain the choices you made.

6
Provisioning and Implementing an Azure Cosmos DB Database

After learning about big data, clusters, partitioning, and setting up JSON structures, it is time to put it all together and actually create a Cosmos DB database and a container to actually store the data. We will see how to implement partitioning. We will then cover some basics of working with the database created.

In this chapter, we will discuss the following topics:

- Provisioning a Cosmos DB database
- Creating a container
- Uploading documents to a container
- Importing data using the Azure Cosmos DB Data Migration tool

Technical requirements

You will need the following requirements if you wish to successfully complete this chapter:

- You will need a connection to the internet and a modern browser.

- You will need an Azure subscription with permission to create new services.

- You will need permission and the ability to download and install the Azure Cosmos DB Data Migration tool.

Provisioning a Cosmos DB database

As we did in *Chapter 4, Provisioning and Implementing an Azure SQL Database*, to create a database, we will go through the process of creating a Cosmos DB database step by step. Different settings are described in different sub-sections, but all steps should be followed (although you could create a database with default settings after entering the settings on the first page with basic settings only).

Let's get started:

1. Start a browser and open the Azure portal (`portal.azure.com`).

2. Open the portal menu (click on the button with three horizontal lines in the upper-left corner of the portal home page).

3. Click in the menu on the **Create a resource** button.

4. Type `Azure Cosmos DB` in the textbox at the top of the page to search the marketplace.

5. Click on the tile for **Azure Cosmos DB**.

6. Click on the **Create** button.

You now get to a page with multiple tiles where you need to choose the type of API you want to use. Cosmos DB is said to be a multi-model database server. It can be used as a document database, a wide columnstore database, a key-value database, or a graph database. When used as a document database, you use either the Core (SQL) API or the Azure Cosmos DB for MongoDB API. You use Cassandra, Azure Table, or Gremlin in the other cases, respectively. These APIs make migrations from older implementations using Cassandra, Azure Table, or Gremlin to Azure easier. The same goes for the MongoDB API, which facilitates migrations from MongoDB to Cosmos DB. For new implementations, the Core (SQL) API is recommended. Using this API, we can write SQL queries to read data from our Cosmos DB document database:

- Click on the **Create** button in the tile that reads **Core (SQL) – Recommended**.

A page with multiple tabs opens, as shown in *Figure 6.1*:

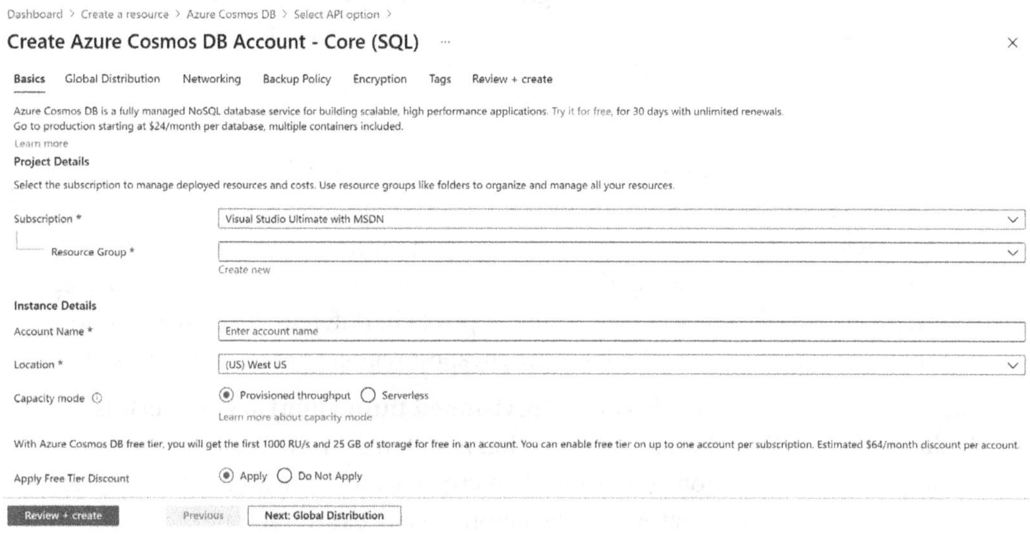

Figure 6.1 – Create Cosmos DB Basics page

There are a couple of settings you need to provide, very similar to the way you created an Azure SQL database in *Chapter 4, Provisioning and Implementing an Azure SQL Database*. We will go through all the settings in the following sub-sections.

Basics

On the **Basics** tab, we find four settings that we need to configure. These settings are as follows:

- Providing the subscription
- Providing the resource group
- Choosing the API
- Choosing between provisioned throughput and serverless

The first two settings are the same for every new resource you create. You provide the subscription to which this resource will be billed. You also provide (or create) the resource group you want your Cosmos DB database to be part of:

1. Select the subscription you want to use and select the **DesignDatabase** resource group you created in *Chapter 4, Provisioning and Implementing an Azure SQL Database*.

2. Type `cosmosdb-book` as the account name.

    A Cosmos DB account name must be globally unique, like an Azure SQL database server name. It forms the entry point to your database. You may not use capital letters, and most special characters are not allowed.

3. Select the **Location** closest to you in which to create your Cosmos DB account.

    When choosing the location, the same arguments apply as when creating an Azure SQL database. The most important aspect is to keep the application and database close to each other to reduce latency. Also, keep in mind that you pay for data leaving a data center. There might be legal requirements regarding where you keep your data. And lastly, services have different prices in different Azure regions, so you might save money if you look for the cheapest region to create resources in.

    The next setting is choosing between **Provisioned throughput** and **Serverless**. At the time of writing, the **Serverless** option is still in preview. Again, this choice is comparable to a decision you made when creating an Azure SQL database. The serverless option means that Azure will automatically use the amount of resources that is necessary based on the current workload. Azure will scale up and scale down as needed. With a varying or unpredictable workload, this is the cheapest option.

    When your database has a constant workload and you know what resources your database needs, provisioned throughput is cheaper. You will provision resources later on a per-container basis.

4. Keep **Provisioned throughput** selected.

5. For now, keep `Apply` selected for **Apply Free Tier Discount** and `Non-Production` selected for **Account Type**.

6. Click **Next: Global Distribution** to move to the next page of the creation process.

7. Keep the `Disable` option selected for the three options, **Geo-Redundancy**, **Multi-region Writes**, and **Availability Zones**.

    When you enable geo-redundancy, a replica of your Cosmos DB database will be created in the region paired to the region where you create your Cosmos DB database. This will increase the availability of your database. When one region is down, you should still be able to connect to the replica in the other region. You can enable this for a database at a later point in time. You can also add more regions to your account later.

When you have multiple replicas of your database in different regions, you can configure one database to be the master that you write to. You can always read from all replicas. The idea is that a global application runs in a data center that's closest to the logged-in user and uses a Cosmos DB replica in the same region as the application runs. This reduces latency and increases performance.

With multi-region writes enabled, the application can not only read from the closest replica, but it can now also write to that replica. This also reduces latency for write operations. You will have to handle write conflicts yourself. Write conflicts occur when the same data is altered simultaneously in two different regions. You end up with an original value and two new values, one from each data center where the data was altered. You have to decide which of the three values will be the new consistent state of the database by writing a conflict resolution function.

The Availability zones options spread your database over multiple availability zones, further increasing the availability of the database, as with SQL databases and storage accounts.

8. Click on the **Next: Networking** button to go to the next page.

Networking

Applications can connect to a Cosmos DB database through a public or a private endpoint of the database. With a private endpoint, your database gets a private IP address from a VNet. All traffic to and from the database is now handled over a private link. Only services and applications that are part of the same VNet as the database can connect to the database. This is the most secure option and should be used when your application can be part of the same VNet. A private endpoint is an Azure resource that needs to be created. VNets are also Azure resources that need to be created.

Instead of using a private endpoint, Cosmos DB can also use a public endpoint. In that case, a public IP address is assigned to the database. A public endpoint is accessible through the internet. Using a public endpoint still allows you to select a list of networks that you allow connections from, denying access from other networks. You can also limit connections to IP addresses that you put in an IP address whitelist:

1. For this book, select `Public endpoint (selected networks)` for the **Connectivity method** field.

2. Select `Allow` for the **Allow access from Azure Portal** and **Allow access from my IP** options. (Note that if you continue working with this database at a later date, you may have another IP address that you then have to allow access to as well.)

3. Leave the **Virtual Network** setting empty.

4. Click on the **Next: Backup Policy** button to go to the next page.

Even though you could have clicked on **Review and create**, it is a good idea to quickly look at the backup settings. The backup settings are described in the next sub-section.

Backup policy

At the time of writing, Microsoft is testing a continuous backup policy with a limited set of Cosmos DB customers. For all other Cosmos DB users, a periodic backup policy is automatically in place. You get to configure what periodic exactly means to you.

By default, a backup will be taken every 240 minutes (4 hours), and the backup will be held for 8 hours. This means that two copies will be held. Depending on the combination of your backup interval and the backup retention, you can hold more than two backups. The first two are free of charge. You pay for every extra backup held for you.

For our current situation, the default settings are OK. We will keep all the default settings on this page and go straight to the next page with encryption settings.

Click on the **Next: Encryption** button to go to the next page.

With the backup settings in place, it is time to discuss the encryption settings. We will do so in the next sub-section.

Encryption

Your data stored in Cosmos DB will always be encrypted at rest. This means that data written to a hard disk is encrypted. You can provide your own key or you can let Cosmos DB provide a key. When you use your own key, you have to store it in Azure Key Vault and provide the URL to the key vault:

1. Keep `Service-managed key` for the **Data Encryption** setting.

2. Click on the **Next: Tags** button to go to the next page.

3. Click on the **Next: Review + create** button to go to the next page (the tags are the same as when creating an Azure SQL database).

4. Click on the **Create** button to create the Cosmos DB account.

5. Click on **Go to resource** once Azure has created your Cosmos DB account.

Notice that you just created an account. This is still free. The next step will be to create a database and to create collections within that database.

Creating a container

The easiest way to create a database and a container is by using the **Data Explorer** page in the portal:

1. Click on the home blade of your Cosmos DB account in the menu on the left-hand side of the screen, and then click on the **Data Explorer** option.

    Notice that you can create a sample database to familiarize yourself with Cosmos DB by clicking on **Start with Sample** in the middle of the screen. We will create our own database with our own container in it. When you create a container without first creating a database, the portal will let you create the database as you go. A database is a container that holds a set of containers, in the same way that a SQL database has a set of tables. You can create the database first.

2. Click on the **New Database** link or click on **New Container** in the menu in the top-left corner of the **Data Explorer** page, and then click on **New Database**.

3. Name your database `ProductDB`.

    Under the database name is the **Provision throughput** option. Throughput is calculated with **Request Units per second (RU/s)**, as discussed in *Chapter 5, Designing a NoSQL Database*. You can provision the throughput at the database level, or you can assign request units at the container level. Request units assigned at the database level are shared among all containers in the database. This does not provide you with any guarantee at the container level. Multi-tenant applications, where each tenant gets their own container, are an example of when you might want to provision resources at the database level. You can compare it to elastic pools in Azure SQL databases, where you don't assign resources per database but per group (pool) of databases.

    Also comparable to Azure SQL databases is the choice between **Autoscale** and **Manual**. Using **Autoscale**, you set the maximum request units you want to use (pay for). Cosmos DB will monitor your workload and assign as many resources as necessary, up to the maximum specified, based on the current workload. This is a good option for databases with varying workloads. For databases with a steady workload, the **Manual** option is cheaper.

4. Deselect **Provision throughput** and click on the **OK** button.

    Notice that on the left-hand side of Data Explorer, you can find the database you just created under **SQL API**.

5. Hover with your mouse over the **ProductDB** database and click on the three dots that appear behind the name.

6. Click on **New Container** in the pop-up menu.

7. Check that the **ProductDB** database is selected in the dropdown under the **Database ID** setting.

8. Name the new container `Product`. (Notice that they use the term *Database ID* for its name.)

   As we saw in *Chapter 5*, *Designing a NoSQL Database*, choosing the right partition key is really important in setting up a container. We will store documents as shown in the following code block:

```
{
    "id": "1",
    "productid": "33218896",
    "category": "Women's Clothing",
    "manufacturer": "Contoso Sport",
    "description": "Quick dry crew neck t-shirt",
    "price": "14.99",
    "shipping": {
        "weight": 1,
        "dimensions": {
        "width": 6,
        "height": 8,
        "depth": 1
        }
    }
}
```

   The **productid** item is the best option to use as the partition key (assuming it has the most distinct value and will be used by the application to retrieve product information). The partition key should always start with a slash. This denotes the root. With embedded objects, the partition key might be an item stored within an embedded object. The item name should then be prefixed with the object name. For example, if you want to use **weight** as the partition key, you should type `/shipping/weight` as the partition key to denote that weight can be found in the nested **shipping** object.

9. Type `/productid` in the textbox under **Partition key** to make the product ID the partition key.

Notice that you have to tell Cosmos DB when your key is larger than 100 bytes.

As you can see, you can set the throughput you need at the container level. The request units you specify here are guaranteed for this container. Specifying the request units per container provides you with more control over the performance of this specific container. This means that specifying the request units per container is, in most scenarios, preferable over setting them at the database level.

Notice that there is a link to a capacity planner. This is a tool that helps you to estimate how many request units you need. You would normally start small and increase the number of request units later when needed to avoid wasting money. As we have already said, choose **Autoscale** for varying workloads.

The last setting is where you choose to expose the data in this container to Azure Synapse Analytics. Cosmos DB is optimized for OLTP workloads. It is an operational database, not an analytical database. But you probably need the data stored in Cosmos DB for BI as well. You can do normal **Extract Transform Load (ETL)** to get data out of Cosmos DB and into a data warehouse or data mart. These databases are then used by BI to create reports and dashboards or to do more advanced analytics.

You can also set up an Azure Synapse Link in your Cosmos DB account. After setting up the Azure Synapse Link, you can add containers to the analytical store. This means that Cosmos DB automatically creates a copy of the data stored in the container in a columnstore structure. A columnstore structure is a structure optimized for analytics (for OLAP workloads). This data is now available from Azure Synapse Analytics. You can now get real-time insights into your operational data stored in Cosmos DB without having to write (possibly complex) ETL processes yourself.

10. Choose **Manual** throughput with 400 request units.

11. Click on the **OK** button to create the container.

You can create multiple containers per database and multiple databases per account. Within a container, you store your data. The data is stored as JSON documents. Depending on the API you choose, different terminology might be used. You can see the structure in *Figure 6.2*:

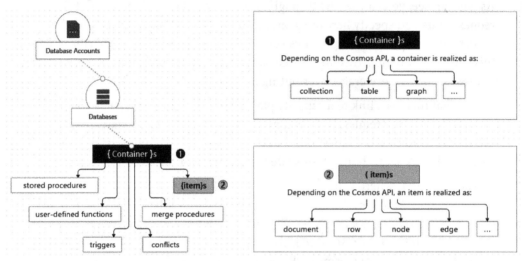

Figure 6.2 – Cosmos DB structure

As you can see in *Figure 6.2*, a container contains items but also stored procedures, user-defined functions, triggers, conflict resolution, and merge procedures. It is outside the scope of this book to go into the other objects. We will now add some data to our container.

Uploading documents to a container

Now that we have a container, let's upload the JSON document from the listing in the previous section. You can find this document and a second similar document in the downloads with this book. They are called `ProductInfo 1.json` and `ProductInfo 2.json`.

After adding a container to a database, the container is listed under the database on the left-hand side of the **Data Explorer** page:

1. Click on the **Product** container to open the container.

2. Click on **Items**.

3. Click on **Upload item** and browse to the downloads for this book. Select both `ProductInfo 1.json` and `ProductInfo 2.json` and click on the **Open** button.

4. Click on the **Upload** button.

Your screen should now look similar to *Figure 6.3*:

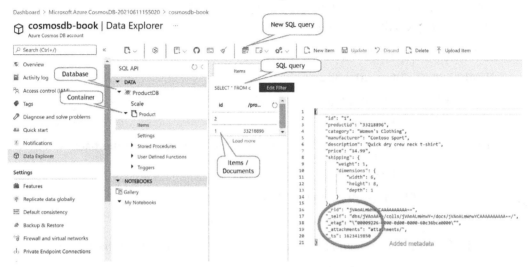

Figure 6.3 – Cosmos DB Data Explorer

After uploading a document, it will be added to the **Items** folder in the container. By default, a pane is opened that shows a list of items in the container. When you click on an item in the list, the document is shown on the right-hand side. Notice that Cosmos DB has added metadata to the document.

Once you have data, you can start querying the data.

5. Click on the **New SQL query** button (see *Figure 6.3*).

Notice that the following SQL statement is already there on the new page (or the new tab) that opens:

```
SELECT * FROM c
```

6. Click on **Execute Query**.

Notice that you get both documents back, including the extra added metadata columns. One oddity in the SELECT statement is the c in FROM c. Cosmos DB wants you to use SQL, and within SQL you should specify the table you read from in the FROM clause. To keep the statements the same, you need a FROM clause here as well, but it doesn't really mean anything. You are already connected to the container you are reading from.

7. Change the SQL statement to `SELECT * FROM c` and click on **Execute Query**.

    Nothing changes because the name used after the `FROM` keyword is just a placeholder. This changes slightly when you realize that the JSON might contain embedded objects that you want to query.

8. Execute the following query:

    `SELECT * FROM p.shipping.`

    This query gets the JSON object shipping from all documents that contain such an object. Make sure to write the casing correctly because Cosmos DB is case-sensitive.

9. Try the following statements and notice that you get individual values back:

    `SELECT * FROM p.productid.`

    `SELECT * FROM p.shipping.weight.`

    You can use a `WHERE` clause to search using specific criteria.

10. Try the following statement, and notice you only get one document back:

    `SELECT * FROM p WHERE p.productid = "33218897".`

11. To retrieve a list of all product IDs that have shipping information stored in the JSON document, you could use the following:

    `SELECT p.productid FROM p JOIN p.shipping.`

It must be clear by now that you can use SQL-like queries to get data from your Cosmos DB container.

Cosmos DB container settings

When you query a container, like the examples in the previous section, you want your container to be indexed to ensure correct performance. By default, the indexing is set to automatic. You can adjust the indexing policy to change the types of indexes being created. Indexes require extra writes when inserting or updating data in your container. That means that you need more request units for the same workload when writing. So, indexes come at a cost. Because indexes help to locate data quickly, they enhance the read performance:

1. Open the **Storage Explorer** page.

2. Open the **Product** container.

3. Click on **Scale & Settings**.

Notice the first tab, **Scale**, is where you find the provisioned throughput. This is also where you can switch from **Manual** to **Autoscale** or back.

The second tab, **Settings**, lets you configure **Time to Live**. The Time to Live feature on Cosmos DB will automatically delete documents after an expiry period. Session data for a website might no longer hold any value after a user has logged out. With Time to Live, this type of data can be removed automatically. You don't have to write your own code. Plus, Cosmos DB will use leftover request units to do the work.

The **Geospatial Configuration** option determines whether geospatial data is indexed as **Geography** or **Geometry** data. The first considers the earth to be a sphere, the second a flat space. When you want to calculate the distance an airplane needs to fly to go from Amsterdam to New York, you use **Geography**. If you have created a map of your office building, use **Geometry**.

The third tab, **Indexing Policy**, shows the indexing policy. The **IndexingMode** property can be either `none` or `consistent`. Cosmos DB will automatically create range indexes on all properties it finds in your documents. With the **includePaths** and **excludePaths** properties, you can tell Cosmos DB explicitly which indexes to make and not to make. You can use these properties as well to create composite indexes besides range indexes. An in-depth discussion on indexing is beyond the scope of this book.

Importing data using the Azure Cosmos DB Data Migration tool

Uploading or entering documents manually is nice for a quick test to see whether your container works as expected. One way to easily get started with some real data is the Azure Cosmos DB Data Migration tool. Let's move the **Product** table from the Azure SQL Database called `Northwind` that we created in *Chapter 4, Provisioning and Implementing an Azure SQL Database*, to Cosmos DB:

1. Open a browser and browse to `https://azure.microsoft.com/en-us/updates/documentdb-data-migration-tool/`.

2. Click on the link to the Microsoft Download Center and download the file `dt-1.7.zip` from there.

3. Extract the .zip file.

4. Start the Azure Cosmos DB Data Migration tool by double-clicking `dtui.exe` in the downloaded files.

5. Click on the **Next** button on the **Welcome** page.

6. On the **Specify source information** page, select **SQL** from the drop-down list under **Import from**.

7. Open the Azure portal and browse to the `Northwind` database we created in *Chapter 4, Provisioning and Implementing an Azure SQL Database.*

8. Click on **Show database connection strings** on the **Overview** blade.

9. Copy the connection string found under **ADO.Net** to your clipboard.

10. Switch back to the DocumentDB Data Migration tool and paste the connection string into the **Connection String** textbox.

11. Find `{your_password}` in the connection string and replace it with the password you used when creating the `Northwind` database.

12. Click on the **Verify** button to verify that you can make a connection to your Azure SQL database.

13. Use the following query to read all the products from the **dbo.Product** table:

```
SELECT * FROM dbo.Product;
```

14. Click on **Next**.

15. Switch back to the Azure portal and browse to **cosmosdb-book**.

16. Find **Primary Connection String** on the **Keys** blade from your Cosmos DB and copy it to your clipboard.

17. Paste the connection string to the **Connection String** textbox on the **Specify target information** page of the migration tool.

18. Add `Database=ProductDB` to the end of the connection string and verify that you can connect to your Cosmos DB.

19. Type `Northwind_Product` for **Collection**.

20. Type `/ProductID` for **Partition Key**.

21. Click on **Next** twice.

22. Click on **Import**.

    To see the result of the import, we will execute two simple queries.

23. Go to your Cosmos DB database in the portal and open **Storage Explorer**.

24. Verify that you have a new container called `Northwind_Product`.

25. Open a new SQL query and execute the following two statements:

```
SELECT * FROM c WHERE c.ProductName = "Chai"
SELECT COUNT(1) AS Count, c.CategoryID  FROM c GROUP BY
c.CategoryID
```

Congratulations! You have created a Cosmos DB database. You have created a container inside that database. You manually entered some data into the container, and you used the Cosmos DB Data Migration tool to enter data coming from a SQL database.

Summary

There is a lot more to know about Azure Cosmos DB than we have seen in this chapter. The basis for successful implementation was made in *Chapter 5, Designing a NoSQL Database*, which was on how to choose your JSON documents and how to choose a partition key.

In this chapter, you learned to configure the options needed to provision a Cosmos DB account. Then you learned how to create a database and a container. You added some data to the containers and queried the test data using SQL.

With this chapter, we end part 1 of this book. *Part 1, Operational/OLTP databases*, was about databases that are used by line-of-business applications, such as online shops, CRM systems, and financial systems. These systems should be able to handle new data well. New data gets inserted into these types of databases and possibly at a high rate. We need to build these databases with performance and scalability in mind. Of course, we need to consider cost as well. The first step in that is choosing the right type of database: SQL or NoSQL. The next step is designing the schema correctly. This ensures the best possible performance and scalability in a cost-effective way.

The next part of the book will be about analytical systems, systems that are used to get the most information out of the data we have. Again, the focus will be on designing a database to get to the best performance, but for a BI and analytical workload instead of for an operational workload. The next chapter, *Chapter 7, Dimensional Modeling*, will go into designing a star schema data model that optimizes a relational database for analytical workloads.

Section 2 – Analytics with a Data Lake and Data Warehouse

Analytics requires a different approach to storing data because of the different usage patterns. In this section, we will focus on data stores optimized for analytics.

This section comprises the following chapter:

- *Chapter 7, Dimensional Modeling*
- *Chapter 8, Provisioning and Implementing an Azure Synapse SQL Pool*
- *Chapter 9, Data Vault Modeling*
- *Chapter 10, Designing and Implementing a Data Lake Using Azure Storage*

7
Dimensional Modeling

Normalizing data is not always the best strategy when designing a relational database. We already mentioned several times that normalizing data is beneficial for an OLTP workload. OLTP workloads are workloads of primary processes, that is, of line-of-business processes.

Databases normalized to the third normal form turned out to be bad for query performance when we started doing more analytical queries on the data. Dimensional modeling came up as an alternative method for designing database table structures. Dimensional modeling leads to a database design optimized for analytics. For instance, the resulting star schema is the ideal table structure for Power BI.

This chapter is all about dimensional modeling and the resulting star schemas. We will learn about the following topics:

- Background to dimensional modeling
- Steps to get to a star schema database model
- Designing dimension tables
- Designing fact tables
- Using a Kimball data warehouse versus data marts

Background to dimensional modeling

When relational databases were introduced in the early 80s, businesses were promised that they would never lack information again. Slogans such as "information at your fingertips" and "always make decisions based on facts, not on gut feeling" were used to sell relational databases. These slogans are still used today. This time, they are used to sell business intelligence.

Gartner defines business intelligence as follows (`www.gartner.com/it-glossary/business-intelligence-bi`):

> *Analytics and business intelligence (ABI) is an umbrella term that includes the applications, infrastructure and tools, and best practices that enable access to, and analysis of, information to improve and optimize decisions and performance.*

A more pragmatic formulation could be: provide the right people with the right information at the right time in the right format.

Using SQL, this promise seemed within reach. Using SQL, you can formulate any question (query) and execute that query in real time on the database. Keeping all your data in relational databases then allows you to ask any questions. However, the real world turned out to be more complex than anticipated. To deal with the issues businesses faced, Ralph Kimball invented dimensional modeling:

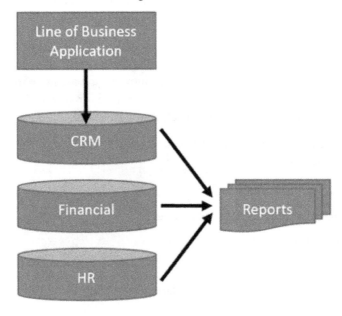

Figure 7.1 – Original architecture

Figure 7.1 shows the original intention. Different departments use different applications that use different databases. Reporting tools connect to those databases and use the data in them to create reports and dashboards. Those reports and dashboards provide the users with the promised information that they can now use to make informed decisions.

It turns out that reporting directly on these operational databases has a couple of shortcomings. The issues businesses are faced with are as follows:

- Performance
- Consistency
- Data quality
- The complexity of normalized database schemas
- Lack of historical data

Let's look at each of these issues in detail.

Performance

We introduced the difference between OLTP and OLAP in *Chapter 1*, *Introduction to Databases*. Normalized databases are well suited for an OLTP workload. This is the type of workload line-of-business applications generate. Reports generate the type of queries categorized under OLAP. OLAP workloads have the following two characteristics:

- The workload is (almost) read-only.
- Most queries need large datasets (lots of rows) to generate the requested information.

Part of what normalizing does is optimize the write workload to a database. This is not beneficial for reading. Normalizing data does optimize the writing by dividing columns over multiple tables. The side effect is that when you need to create a report, you have to get the requisite data from multiple (possibly lots of) different tables. This leads to lots of joins. When many joins are needed, the complexity goes up and the performance goes down.

Another consideration to make is that running a report on a line-of-business database has an impact on that database. Can you afford that your website is slow because someone in the analytics department is running some queries? You need business intelligence to have no (or just a minimal) impact on operational processes.

When you copy data to another database, you can run analytics queries on that copy. Using a different database means you don't have an impact on the original database when executing queries. It also provides you with the opportunity to use a different table structure, that is, a different database schema. You can use a database schema better suited for your intended use: analytical queries.

Consistency

Figure 7.1 shows an example with three different line-of-business databases applications. Some reports might need data coming from different databases. Other information might be found in multiple databases. For instance, you might get sales figures to look at invoices in your financial system. You may get the same information by looking at orders from your CRM system. In real life, both ways of getting sales data will probably give you different numbers. This means that people using information coming from different systems have different information. How can they make informed decisions together when they don't agree on (or don't know what is) the real truth? There is a lack of consistency across the databases. But dashboards and reports need to be able to provide a 360-degree view of your business. They need all the data to be integrated.

Storing all data from all systems in a single database will solve this problem for the most part. When you copy data from the line-of-business databases to a new database, you can define which numbers are right and describe the real world best. When everybody queries the same database, that is, when they all use a single source of information, everybody has the same information to base decisions on. Such a central database is called a data warehouse.

Data quality

Lack of consistency across databases results in the bad quality of the information provided within reports. Data of bad quality leads, likewise, to bad reports. For instance, say you use Power BI and drag the City column from the Customer table to the report followed by the CustomerID column. You configure Power BI to count the number of CustomerID columns so that you have the number of customers per city in the report. Are the numbers shown correct?

What if you have customers who live in New York, but you also have customers for whom the City column holds the value New York City? The database also has the value NY city. These three different values are synonyms. They all mean the same city. You might have typos in city names as well, for instance, newyork. Power BI just sees different values. If your query was to show the top 10 cities based on the number of customers, New York might not be on the list because you have these different variations on how it is written in the database.

Another common data issue is incomplete data. The Blank, Empty, or NULL city might make it into your top 10 list. As a final example, you will probably have multiple rows for what is actually the same customer, for instance, someone who forgot their password and created a new account instead of resetting the password.

In other words: real, simple, straightforward analysis can lead to wrong results because of a number of data quality issues present in the line-of-business databases. Data modeling is not going to help too much in solving the previously mentioned data quality issues. You can, however, try to correct the issues before loading the data to a new database: the data warehouse.

The complexity of normalized database schemas

Consider a self-service BI scenario where people use Power BI to connect to their database and start creating visualizations using drag and drop. This might work out just fine with a database such as Northwind. Northwind has eight tables with clear names for both the tables and the columns. The sales process behind the data is also straightforward. But what if the database has thousands of tables with many columns per table and both tables have non-descriptive names? Will drag and drop provide the insight you are after or do you need to write complex DAX expressions? Will Power BI still perform well with filters being applied over many relationships? Or do we need a database schema optimized for tools such as Power BI? The answer to the last question is, of course, yes!

Even when self-service BI is not an argument, complexity is still an issue. Experienced BI developers might be able to write the needed SQL queries or DAX expressions. But even experienced professionals make mistakes. Complex expressions tend to be error-prone and bad for performance. Besides that, developers are more productive when the complexity goes down. Star schema databases are simple and intuitive and, as a result, address the issues mentioned here. Creating a new database modeled as a star schema helps.

Lack of historical data

Lastly, the analysis that we can perform on data from line-of-business databases is often limited by the fact that a lot of databases do not keep track of historical data. Historical data means data such as, for instance, the address information of customers. We tend to never throw away rows. So, all orders ever stored in the database are still there and can be used for querying. But when a customer moves, a lot of databases tend to overwrite the old address with the new address. By doing so, you throw away the old address.

Suppose you want to do an analysis to figure out which products are popular in which neighborhoods. You might do more targeted marketing or adjust product prices when you know exactly what kinds of products are bought more or less often in a certain neighborhood. The problem is that in order to do the requested analysis, you need to know where a customer lived when they purchased the product. They might have moved since then. This means you do not have the data required for the analysis.

Creating a data warehouse can also solve this last issue. Whenever you detect a change in the address of a customer, you add the new address while keeping the old address. A data warehouse is a database where you store historical data in a schema optimized for analytics. When you can load good-quality data into the data warehouse, all of the previously mentioned issues you face when doing analytics on line-of-business databases are addressed. The architecture of *Figure 7.1* changes into the architecture shown in *Figure 7.2*:

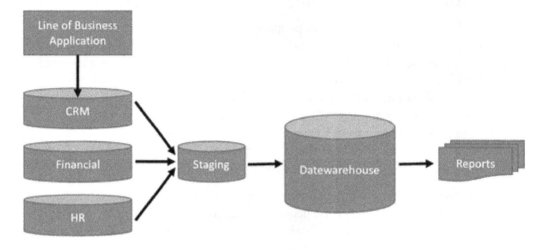

Figure 7.2 – Classic data warehouse architecture

We will get to more modern architectures when discussing Data Vault, and especially when covering data lakes. The data warehouse depicted in *Figure 7.2* can and should be optimized for its intended use: performing analytics that generate an OLAP workload. It makes sense to design a data warehouse in a different way to how you design your operational database. An office building is designed differently from a school building. They are both buildings but they are used in different ways.

The database called staging that you can see in *Figure 7.2*, between the operational databases and the data warehouse, breaks the action of copying data to the data warehouse into two main pieces:

1. We first extract the data from the operational database and store the exported data.

2. We then cleanse the data and transform it before loading the data into the data warehouse.

The entire process of **extracting, transforming, and loading** data is called the **ETL** process. You will learn more about the ETL process in *Chapter 11, Implementing ETL Using Azure Data Factory*.

The conclusion so far is that we face some issues when performing analytics on normalized databases. As with Cosmos DB, we should adjust our design to be a better fit with the intended queries we will run on this database. Dimensional modeling leads to that better fit.

Understanding dimensional modeling

To avoid any confusion, dimensional modeling is a data modeling technique. It leads to a table structure for relational databases. The phrase "table structure" says it all. We are going to store data in tables. We could use Azure SQL Database to implement this database. We will use Azure Synapse Analytics in *Chapter 8, Provisioning and Implementing an Azure Synapse SQL Pool*, to implement the database. Dimensional modeling, in other words, is an alternative to normalizing data. It will use relationships between tables based on primary keys and foreign keys.

Let's briefly reiterate the main principles we had for normalizing data and look at why they are not helpful when creating a database that should be optimized for analytics. These main principles involved in normalizing data are as follows:

- Minimize redundancy.
- Use dependencies between attributes.

Minimizing redundancy

Minimizing redundancy makes writing into the database faster. A data warehouse, however, is optimized for read performance, making this last argument invalid. De-normalizing tables can increase the read speed significantly.

Minimizing redundancy also means that you minimize the needed storage. However, storage costs are low nowadays. But even more importantly, when you are working with large datasets, such as in data warehouses where we combine data from different systems and where we keep track of historical data, the databases are large anyway. The extra storage required by de-normalizing is not that much, relatively speaking. We choose performance over this little extra storage, saving time and the necessary compute instead of saving on storage.

Using dependencies between attributes

The second principle of normalizing data is to use the dependencies between atomic data to model the database. For instance, price and weight are characteristics that depend on a specific product. Therefore, they are columns in a `Product` table. The values depend on the value of the key that defines the table.

Dimensional modeling does not take the dependencies between columns as its primary starting point. Dimensional modeling models the business processes, not so much the data itself. Making the processes central to the design means making the information needed for these processes important rather than the underlying data. As we saw earlier, we adopt the intended usage and make that the lead in the design. That will lead to better read performance.

This is, by the way, true for normalizing data as well. The important thing with normalizing is that a big part of the intended use is that we register what we do, making entering new data into the database really important. So important, in fact, that we optimize the database largely to that aspect of its intended usage. Dimensional modeling optimizes a database for read performance.

The result of dimensional modeling is a so-called star schema. Let's look at what that looks like next.

Understanding star schemas

Dimensional modeling makes a distinction between two different types of tables. These two tables are the following:

- Fact tables
- Dimension tables

The distinction between fact tables and dimension tables is a logical distinction only. A table is a table. A table consists of columns with a name and a data type. A table can have a primary key and we use foreign keys to relate tables to one another. Technically, there is no distinction between the two different table types. Logically, however, there is.

The premise of dimensional modeling is that we create a star schema design for each process in an organization that we want to create reports and dashboards for. The process itself is the starting point of our modeling journey. A star schema describing the process is the end point. You can see an example of a star schema in *Figure 7.3*:

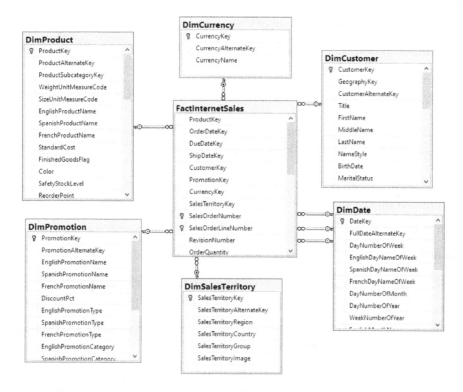

Figure 7.3 – Star schema

Let's look at fact and dimension tables more closely.

Understanding fact tables

Central to the star schema is the fact table. The fact table stores all the measurable quantities that enable us to quantify a process. These quantities are called facts or measures. These are the main values we are interested in on reports. These are the values our analysis efforts focus on. Let's take a webshop that sells products to individual customers as an example.

The sales process of our exemplary webshop can be made measurable by looking, for instance, at the sales amount. The sales amount is a **fact**. It is an objective number. It is the registration of an actual amount that switched hands. We can compare the amount to a target or an expectation to give it a more subjective value, such as "good" or "bad," because the amount is higher or lower than the target. We could also compare it to the amount of another, but comparable, period, such as comparing the sales of January 2021 to the sales of January 2020.

Other examples of facts could be the following:

- The number of products sold

- The margin that the webshop made on a sale

- The discount they gave to lure the customer into buying the product

Other processes will have different facts because we quantify them in another way. Examples of different facts include the following:

- Bed occupancy in a hotel or hospital

- The number of days that employees call in sick for human resources

- Hours worked in consultancy

- The number of students who have passed exams

Every number that tells you something about whether a process is doing well can be seen as a fact.

In dimensional modeling, we store all the facts of a single process together in the fact table. Each fact becomes a column in the fact table.

A fact is almost always a numerical column. Numerical data is aggregatable. You can, for instance, add the sales amounts of all orders placed in the same month together to form the total sales of that month. Likewise, you could add all the sales amounts of orders of a specific customer to get the total amount that the customer spent in your shop. You can use ways to aggregate data other than the mentioned summing of amounts. You can calculate an average sales amount, for instance, or you can count the number of unique customers visiting your website. Of course, you can count everything, independent of the data types. Most aggregate functions, however, need numerical data.

Facts are also called **measures**. They make the process measurable. Strictly speaking, a measure is an aggregated fact. The individual sales amounts of individual transactions are facts. The sales, calculated by summing sales amounts, are then the measure.

Facts and **Key Performance Indicators** (**KPIs**) are closely related. KPIs are used to track how well companies are doing.

> **Note**
> A KPI is a management tool to show the status of a specific process.

A KPI consists of a fact, an objective numerical piece of data, such as, for instance, sales. This fact is linked to a target. How high should sales be for our business to be able to say whether we are successful? A target could, for instance, be that we want to grow by 10% compared to the same period in the previous year. The third component of a KPI is the status. The status is an indicator providing a subjective interpretation of the fact based on the comparison of the fact and the target. It is a business rule where you, for instance, say that you are happy when sales exceed the target, you are neutral when you get to 90% of the target, and the business is considered to be very bad when your actual sales are below 90% of the target. KPIs are often shown graphically, for instance, using a traffic light to show the status.

Targets are, in a sense, facts by themselves. You store them in their own fact table.

So, fact tables store facts that quantify business processes. The facts, however, are only numbers that do not have any meaning without context. 1 million, for instance, is just a number that can mean anything. By itself, it means nothing. But when you say that a business has a revenue of 1 million per business unit per month, that value all of a sudden gets a lot of meaning. The words "business unit" and "month" add context (meaning) to the number 1 million.

Business unit and month are examples of dimension attributes. We store dimension attributes in separate tables called dimension tables. The fact table has foreign keys that reference the dimension tables.

> **Note**
> A fact table consists of numerical columns storing the measurable quantities that quantify a process, plus foreign keys that reference the dimension tables, which add context to the facts.

Understanding dimension tables

A star schema consists of a fact table that has many-to-one relationships with a couple of dimension tables. In *Figure 7.3*, you can see the fact table called **factInternetSales** surrounded by dimension tables such as **dimProduct** and **dimCustomer**. A dimension table has a primary key and consists further of columns with descriptive information in them. A dimension table has mainly alphanumerical data in contrast to the numerical data of fact tables.

> **Note**
> Dimension attributes (columns in dimension tables) mainly contain textual descriptors that provide context and meaning to the facts stored in the fact table.

You could say that the dimensions are the most important part of the data. As I already said, a number by itself is meaningless. It is the dimensions that put the facts into context. The context provides meaning to the numbers. We use the descriptive information in dimension tables in three different ways:

- Using query constraints (filters)
- Using group by
- Using report labels

Using query constraints

Suppose a business analyst within a multinational company is doing a sales analysis specifically geared to sales in the Netherlands. All transactions (all facts) need to be filtered on the Netherlands. We only want transactions from the Netherlands to be part of the data we analyze today. We can do this easily when we have a column, that is, a dimension attribute, `Country`, that has `the Netherlands` as one of the values stored in the column. We use the dimension attribute to filter the data.

Using GROUP BY

`GROUP BY` is the SQL `SELECT` clause that allows you to aggregate data. The business analyst of our previous example might want to compare sales data of the Netherlands to surrounding countries. They might then filter on `West Europe` instead of on `the Netherlands` and show sales by country.

By far the most reports in use in business show aggregated data. Dimensions enable us to aggregate data (aggregate the facts) in different ways, such as sales by month, sales by country, and sales by product

Using report labels

We said that dimension attributes provide meaning to facts. We literally see this on reports. Envision a bar chart, for example. On the axis of the bar chart, you will find labels. Labels are descriptive values that tell you what the bars on the chart are about. These descriptive values are stored in a column in a dimension table.

A dimension table can be seen as an important entity. For instance, a simple sales process is about selling a product to a customer at some point in time. This leads to the use of three-dimensional tables. You will have a date dimension that will have a column for each period that you want to use during sales analysis. You might, for instance, have columns for `Year`, `Month`, and `Quarter`. You will also have a product dimension with columns for all attributes that describe a product and a customer dimension with columns that describe a customer. These descriptive data fields are not normalized. With dimensional modeling, we put as many descriptive columns in the appropriate dimension table, even when that leads to redundancy.

In real life, your processes will be more complex than the simple sales process described above. At the same time, that does not necessarily lead to a lot of dimension tables. There is even a best practice stating you should limit the number of dimension tables to a maximum of seven. A star schema with seven dimensions is still clear for an average human being. It starts to be confusing with more than seven dimensions. A key design principle is to make a star schema intuitive and simple to use for people that understand the underlying business process.

Seven dimensions might seem to be limited for describing real-life processes. Dimensions, however, are "flattened" (de-normalized) tables. Flattening, or de-normalizing, means that you store more columns in the same table, meaning you need many fewer tables overall. It might even be unnecessary to have seven dimensions. Less might prove to be enough. Seven is just a theoretical guideline. When a process is complex and you need more than seven dimension tables to describe the process accurately, then so be it. Just keep in mind that reducing the complexity of the database model and increasing the usability from a business user perspective is key to dimensional modeling.

A fact table, together with the dimension tables, is called a star schema. A star schema is a model of a business process; it describes that process. A star schema is the result of dimensional modeling. In other words, dimensional modeling leads to a star schema. Star schemas are the ideal data models for most business intelligence tools, such as, for instance, Power BI.

> **Note**
> A star schema has a central fact table surrounded by dimension tables. This means that the ERD looks like a star. One star schema describes one process.

Let's look at the steps you perform when using dimensional modeling before we look into fact and dimension tables in more detail.

Steps in dimensional modeling

As with normalizing data, we can distinguish some formal steps that lead to a data model. Unlike with normalizing data, these steps are used in an iterative approach. The process of designing the star schema leads to new insights that might mean that you have to go back one step. The steps that you need to take are as follows:

- Choose a process and define the scope.
- Determine the needed grain.

- Determine the dimensions.
- Determine the facts.

Choosing a process and defining the scope

As we already said during Entity Analysis (*Chapter 2, Entity Analysis*), it is easy to get carried away and model the entire world. Make a clear choice in what the goal of the star schema is. Remember that a single star schema models a single process. A business process can be defined as a set of activities and tasks that, once completed, will accomplish an organizational goal. Examples are sales, marketing, human resources, and so on. The ultimate goal of the database is to enable people to monitor the business process, get insight into that business process, and enable them to make decisions to improve the process.

Also, remember that simplicity increases usability and performance. This is especially true in self-service business intelligence scenarios where business users will use Power BI to do analytics with your star schema as the data source.

When you create a star schema to create management reports, you may know exactly what is on those reports beforehand. The scope is now clear. You add what you need and nothing more. The model stays small, intuitive, and easy to use. With self-service analytics, it is slightly more difficult. Your data analysts might want to use everything. The more information you can add to the model, the better you will serve their needs.

Determining the needed grain

The grain is the level of detail of the data itself. The grain defines how much information users can get from this database. It also determines how big the database is going to be. A bigger database is more expensive and probably needs more attention to provide acceptable performance. More detail means increased usability against increased costs.

When you design a star schema to use in Power BI using the **Import** mode to connect to data, size really matters. Power BI is limited in the amount of data it can import. When you create a corporate data warehouse, the usability probably outweighs the cost.

Consider, for example, a supermarket that operates throughout the country. A row is added to the fact table of the data warehouse each time a product is scanned. This is a lot of rows leading to a very big fact table. But afterward, you know exactly which customer bought which product in which store at which time and for which price. That's a lot of detail to analyze.

Alternatively, you could store a row each time a customer pays at the cash register. If customers buy 10 products at the same time on average, you now have just a tenth of the rows in your fact table compared to storing a row for each scanned product. All individual product prices are already added together to form a total transaction amount. The downside is the loss of detail. A simple question such as "How much bread has been sold between 10 and 11 A.M.?" can no longer be answered.

Most of the time, the lack of detail that results from storing data at a high grain is considered to be far worse than the extra costs involved in storing all the details. But if you keep your historical data in a data lake (see *Chapter 10, Designing and Implementing a Data Lake Using Azure Storage*) and you create a star schema in Azure SQL Database or Azure Synapse Analytics, you have a special purpose in mind for this data. Choose the highest grain possible that will allow you to reach the intended goal but minimizes the cost to reach that goal.

Determining the dimensions

When determining the grain, you already determine the main dimensions to use. You could, for instance, create a grain statement such as "We track sales per product, per customer, per day." That is equal to saying that you have dimensions for product, customer, and date. The statement means that when a customer buys bread in the morning and they come to another store in the afternoon to buy more bread, you store those two separate transactions as one row in the fact table. You may need to revise the grain statement to something such as "We track sales per product, per customer, per day, per store." This adds a store dimension.

Changing from "per day" to tracking the time of the purchase also makes sure that both individual transactions are stored independently of each other. This means that you revised the grain.

Adding a transaction dimension with something such as a transaction ID in it will also lead to storing two rows instead of just one.

The primary key of the fact table is made up of the foreign keys of the dimensions referred to in the grain statement. More dimensions lead to more unique combinations of foreign key values, leading to more detail in the fact table.

Not all dimensions need to contribute to the grain. In other words, there might be dimensions that you need to add to the model that are not mentioned in the grain statement. You might, for instance, need to keep track of whether a purchased product was sold with a discount because a special action was running that day. You can add a `SalesAction` dimension to the star schema.

Determining the facts

The last step is to determine the facts. The facts are the numbers used to base business decisions on. What is it your reports need to show?

Determining facts such as `SalesAmount` and `DiscountAmount` is, most of the time, not so hard. There are some things to consider, though. Consider, for example, a company where the margin is a leading factor in their decision making. Suppose the margin is defined as the difference between the selling price and purchase price divided by the selling price. How many measures do we store and which ones?

Even though it is said that margin is important in decision making, it might be better to not store the margin directly in the star schema. The margin, as defined here, is a percentage. Margins are best calculated when you know the context in which they are being used. Adding 12 margins of 12 months to get the margin of the entire year doesn't make any sense. Just adding those values to get to a total is the default behavior of Power BI. So, a Power BI developer will have to overrule this default behavior and define the margin using DAX expressions. With that in mind, store the `SellingPrice` and `PurchasePrice` base values as facts in your fact table and calculate the margin at the time it is needed.

An alternative is to define the margin as an amount in real currency, for instance, in Euros. Adding the margin made in each month to reach a yearly margin now does make sense. A Power BI developer can still show it as a percentage on their report using DAX expressions. This alternative might be easier with both `SellingPrice` and `PurchasePrice` varying over time, which might make it difficult to calculate the margin at a later time than when the actual transaction happens.

Now that we know what a star schema is and we know the high-level steps to design one, let's take a detailed look at what dimension tables look like exactly.

Designing dimensions

The first thing to look at is the primary key to use for a dimension table.

Defining the primary key of a dimension table

To get straight to the point: we always use surrogate keys for dimension tables. In *Chapter 1, Introduction to Databases*, we discussed logical versus surrogate keys. We will not repeat the discussion here. The best practice is to use surrogate keys for dimension tables.

In a star schema database model, using an efficient primary key is even more important than in a normalized OLTP database. In earlier examples, it became clear that fact tables might become really big in terms of the number of rows they store. Suppose you have a fact table with seven dimensions that has 1 billion rows. The difference between using keys that are 4 bytes in size and keys that are 8 bytes in size is 7 x 4 x 1,000,000,000, which is 28 GB. Some people might argue that today 28 GB is not really something to consider. But you might have a lot more rows than 1 billion, you might have more than seven dimensions, and you might have keys even bigger than 8 bytes in size. The extra storage and compute needed to work with the data mean you pay extra for your database in Azure. As an extra argument, the scalability of the database is better if you take large numbers into account even when you do not have that many rows at the time you design the database.

Using `int` as the data type of a primary key allows you to store more than 4 billion unique values. As long as a dimension table has fewer than 4 billion rows, `int` will suffice. `int` is the most efficient type to use.

> **Tip**
> Use a surrogate key as the primary key for dimension tables.

Rows in a star schema always have a source database that they are based on. You always store the primary key of the original row in the source system. This way, you can trace information back to the source when you know the source. That makes it a good idea to add a column with the name of that source system. You might also add a column with a timestamp of the moment a row gets loaded into the data warehouse and a column to indicate which ETL process performed this load.

Adding an unknown member

You will encounter source systems with incomplete information. You might, for instance, find an order in a CRM system for which the account manager responsible for that order is not registered. Something has been sold; there is a fact row describing the actual transaction, but a piece of information is missing.

Whenever you write SQL queries on the source system, you need to be aware of these sorts of situations. If you retrieve a list of all account managers and calculate their respective sales amounts, you might miss all the orders that do not have an account manager. When writing the query, you need to make the distinction between using an `INNER JOIN` command and using an `OUTER JOIN` command. It is easy to make a mistake.

In a star schema, you want to make sure you have a dimension row for each row in the fact table. If your fact table has a foreign key to a `dimAccountManager` table, that foreign key should always have a value. We do not want NULL values in the foreign keys in a fact table. Without NULL values, you can always use INNER JOIN. This makes the star schema less error-prone. It also ensures that tools such as Power BI, where queries are generated by the tool most of the time instead of written by a human being, retrieve the correct data from the database.

To make sure we always have a value for the foreign keys in the fact table, we add a dummy row to each dimension table. We call this row the **unknown member**. We usually reserve the value -1 for the primary key of the unknown member row. Normal rows have key values starting with +1. Each column of the unknown member row gets a value that describes the situation of the missing data. For most alphanumerical columns, this could simply be the text Unknown. For numerical columns, it is less obvious what to use as a value. Simply 0 is the most commonly used value.

With an unknown member row present in dimension tables, the ETL process loading the star schema can use the key value -1 as the foreign key in the fact table each time data is missing. An order without an account manager gets a foreign key value of -1 instead of NULL.

By using the unknown member row in the `dimAccountManager` dimension, you can now retrieve a list of account managers and calculate their respective sales amounts and be sure to always get all sales. There might be an account manager called Unknown in the list. But the corresponding sales amount is showing in the list and the overall sales amount is the correct value.

> **Note**
> Each dimension table should have a dummy row, the unknown member, with a key value of -1 to ensure that each row in the fact table is related to a row in each dimension table.

There might be situations in the previous example where there simply is no account manager for an order. Having no account manager registered doesn't mean the information is missing, but simply means there is an order that no account manager was involved with.

In situations like that, you may need to check the dimensionality of the fact table. Did you define the process and the scope correctly when some orders are apparently different from others? Maybe you have different sales channels that have completely different processes, which forces you to create different fact tables for them. We will discuss this further in the *Designing fact tables* section.

It can be the case that a dimension might not always have a value for each dimension without that meaning that the information is missing. You do not want to relate these rows to the unknown member row. To ensure that each row in the fact table can be linked to a row in the dimension table, you add a second dummy table. This row could get the value -2 as its primary key. The value you add in alphanumerical columns could be something such as Not applicable.

> **Note**
>
> Add a second dummy row to each dimension table with the key value -2 for each fact row for which the dimension is not applicable.

Now that we have learned about keys in dimension tables and two dummy rows to add to dimension tables, let's have a closer look at the design of the tables as a whole.

Creating star schemas versus creating snowflake schemas

We have already mentioned flattening or de-normalizing dimension tables. Take, for instance, a dimension table called dimBook. We add relevant information about books to this one table. *Figure 7.4* shows a simplified version of this dimension table:

dimBook
BookKey
Title
Genre
...
Author
Nationality
DateOfBirth
Sex

BookKey	Title	Genre	...	Author	Nationality	DateOfBirth	Sex
1	ABC	Adventure	...	Peter	Dutch	1-1-1980	Male
2	DEF	Phantasy	...	Peter	Dutch	1-1-1980	Male
...	...	...	...	...	...	...	...

Figure 7.4 – The dimBook dimension table

Notice the redundancy in this table. If an author has written multiple books, all the characteristics of that author, such as their date of birth and sex, are stored multiple times as well. *Figure 7.4* shows twice that an author called Peter is from the Netherlands, was born in 1980, and is male. When normalized, these columns should be part of a separate table called `Author`. In that way, all the information we store about an author is stored only once. In the example of *Figure 7.4*, we flattened the `Author` and `Book` entities into a single table.

The reason for flattening tables is simplicity. All information that business users might use to analyze the facts is just one relationship away from those facts. With a separate table for authors and a fact table that has a foreign key to the `dimBook` table, the `Author` information can only be linked to the fact table by also using the `dimBook` dimension table. Suppose you analyze book sales figures by age category of the author of the book. When you write the SQL `SELECT` statement needed, you have to include the `dimBook` table in your query even though you need no information whatsoever about books. That is not intuitive and takes extra compute power from your database. By flattening the table, you only need two tables and you actually use columns from both of these tables.

> **Note**
>
> In dimensional modeling, intuitivity, simplicity, and query performance are more important than the disadvantages of redundancy. This leads to the use of flattened dimension tables.

A fact table with only directly related dimension tables is called a simple star. There are alternatives to using simple stars, such as, for instance, **snowflake** models.

We speak of snowflakes when one or more dimension tables are normalized instead of flattened. Compare the star schema of *Figure 7.3*, for example, with the data model shown in *Figure 7.5*:

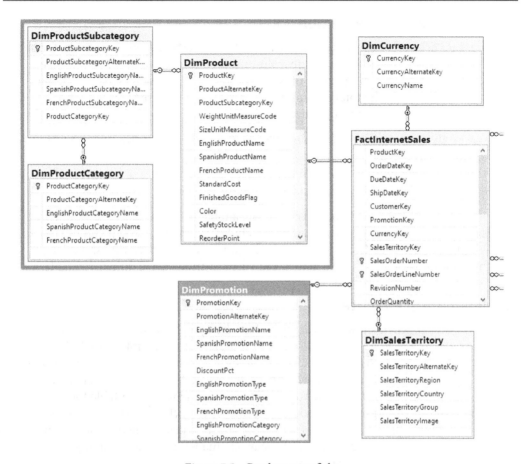

Figure 7.5 – Product snowflake

Look at the **dimProduct** dimension. In the example of *Figure 7.5*, products are categorized into product subcategories and subcategories themselves are further categorized into product categories. There is a one-to-many relationship between product categories and product subcategories. There is also a one-to-many relationship between product subcategories and products. This is modeled in *Figure 7.5* by creating separate tables – **dimProduct**, **dimProductSubcategory**, and **dimProductCategory**. In *Figure 7.3*, all columns from all three tables are combined into the single **dimProduct** table.

The ideal model according to the theory of dimensional modeling is a star schema. Snowflakes are less intuitive and may suffer from slower performance than star schemas. There are, however, two scenarios in which snowflakes may be better or even necessary. These two scenarios are as follows:

- When star schemas lead to monster dimensions
- When you have multiple fact tables that have different grain

Understanding monster dimensions

Flattened tables are much bigger than normalized tables. Flattened tables are wide. Or in other words, flattened tables contain many columns. You don't distribute columns over many tables as with normalizing data, but you put them all in the same table. By doing so, they are also bigger in the total amount of bytes they use because there is a lot of redundant information in flattened tables.

In most cases, the size of dimension tables is negligible compared to the size of the fact table. But when you keep de-normalizing on a dimension table that also contains many rows, the inefficiency might, at some point, outweigh the advantages of flattened tables. A really big dimension is sometimes called a monster dimension.

It is impossible to provide exact guidance on when a table becomes too inefficient. It depends on the actual data, the usage patterns of that data, and the system you use to implement the star schema. Common sense and good performance testing should lead you to the right decision on when to flatten and when to snowflake a dimension table.

Modeling multiple fact tables

The second scenario when snowflakes are useful is when you have more than one fact table in your star schema. We haven't discussed multiple fact tables until now. But suppose the sales team asks you to create a star schema database so that they can make dashboards with sales KPIs. To create KPIs, you need actual sales amounts. You also need to store targets to compare the actuals against. You need to create a separate fact table for the targets. You cannot add targets to the existing `factSales` fact table. There is no target defined for individual orders. Besides, you probably have targets defined before you have actual sales. *Figure 7.6* shows an example of a star schema with two fact tables, one for the actuals of the KPI and one for the targets:

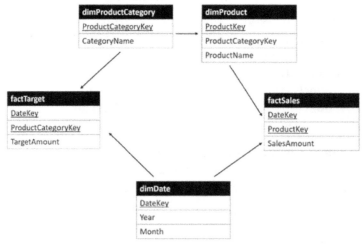

Figure 7.6 – Star schema with two fact tables

We can conclude from the **factTarget** table that targets are defined per category. The **ProductCategoryKey** column is part of its primary key, which means that **TargetAmount** is dependent on **ProductCategory**. If you look at the **factSales** table, you see a **ProductKey** foreign key (and not **ProductCategoryKey**). We sell products but we define targets at the category level. When you roll up (aggregate) all sales transactions from products belonging to the same category, you can compare the aggregated sales amount to the target amount to see whether you achieved your targets.

The problem here is that both fact tables store information that we need to compare to each other, but each fact table has a different grain. The solution is to normalize the `dimProduct` table. In a simple star schema, the `dimProduct` and `dimProductCategory` tables would be flattened into one table, `dimProduct`. All category information would be stored in the `dimProduct` table, but because the `factTarget` table is at a different grain, you cannot link it to the `dimProduct` table. By normalizing `dimProduct` into separate tables for `dimProduct` and `dimProductCategory`, you can maintain the relationship between `dimProduct` and `factSales` and, at the same time, have a relationship between `dimProductCategory` and `factTarget`. Because `dimProduct` and `dimProductCategory` are related as well, you can now aggregate sales to the category level and then compare it to the targets.

Notice that a similar problem might exist for the `dimDate` date table. Oftentimes, targets are defined at the month level, not for individual days. Actual orders are related to an exact date, and not to the month. The `dimDate` table stores a row for every single day on the calendar. You will learn more about the date dimension in the next section. The fact that it has a row for every single day combined with the fact that the `factTarget` table has a row per month means (again) that we cannot relate the two tables to each other directly.

We could normalize the `dimDate` table and create a separate table to hold the months. In this case, however, it is very common to relate targets to the first day of the month. The target for January 2021 gets the value `20210101` as the foreign key linking it to the `dimDate` table. This means we effectively link it to the first day of the month. You just need to make sure that you don't create reports that try to show targets at the day level. As long as all reports are aggregated to the month level or higher, all will work well. Business users understand that they should not drill down to see more detail because they know from understanding their business that such detail does not exist.

The discussion on both monster dimensions and normalizing the date brings us to so-called snowflakes. Let's have a closer look at what that means in the following section.

Understanding snowflakes

A snowflake is a mix of dimensional modeling and normalizing data. It takes the approach that we should build an effective design for whatever we are going to do with the data. That pragmatism is more important than pure theory.

> **Note**
> A snowflake is a dimensionally modeled database where one or more dimensions are normalized.

Snowflakes help with preventing monster dimensions and with relating fact tables with a different grain. They may reduce query performance because extra joins need to be executed and reduce the simplicity and intuitivity of the model. Hybrid approaches exist in an effort to combine the best of star schemas and snowflakes. Let's shortly discuss the following:

- A hybrid model
- Optimal snowflake

Hybrid model

In a hybrid model, attributes are stored in both the main dimension table and in the normalized table. In *Figure 7.6*, that would mean that, for instance, the **CategoryName** attribute is a column in **dimProduct** as well as in **dimProductCategory**.

Using this model, you only need to query two tables when you create a report showing sales by category. That is just one join instead of two. You only need the extra join in cases where you need both the **TargetAmount** column and the **SalesAmount** column.

This approach is not well suited for a Power BI dataset. All DAX expressions need to take into account the fact that a column that looks the same on a report can be a different column from a different table in the model. This likely makes the DAX more complex instead of simpler. When creating a relational data mart, however, this approach could reduce query time and complexity.

Optimal snowflake

An optimal snowflake model adds all the primary keys of all dimension tables to the fact table. In the model of *Figure 7.6*, this would mean that you add the `ProductCategoryKey` key as a foreign key to the `factSales` fact table. The `factSales` table now has a direct relationship with the `dimProductCategory` dimension, ensuring that a report showing sales by category only needs two tables and one join.

The disadvantage of an optimal snowflake is that you increase the size of the fact table. As we have already said, the fact table is the biggest table of the model in terms of the number of rows. Adding a column means storing an extra value for each row. When you use a columnstore structure such as in Power BI or with SQL Server columnstore indexing, this impact depends on the cardinality of the extra foreign key that has been added. With only a few distinct categories, the impact in the example is probably really small.

All four scenarios – simple star, snowflake, hybrid, and optimal snowflake, have their advantages and disadvantages. Start modeling a simple star. Only when you have good arguments for any of the alternatives do you implement the alternative.

By now, you have learned a lot about designing dimensions. Let's now have a look at a special dimension – the date dimension.

Implementing a date dimension

Almost all data analysis involves time in some way. Sales transactions, for instance, take place on a certain day. We use this date to create reports to show sales by month and sales by year and compare the sales amount to the same period in the previous year. We oftentimes use date analysis to do capacity planning. We pay our employees per month and we pay taxes per quarter. Whatever we do, time has a part in it.

Dimensional modeling is about the ease of using the data. Because time plays such an important role in our lives, it is important to make it easy to work with time in our data model as well. You make working with dates and time easy by creating a special date (or time) dimension.

Before you design your date dimension, you look at the grain you need for the fact table. Suppose you want to keep track of sales per day, product, or transaction. Your date dimension should then hold one row for each day on the calendar for the entire period of time you use in your data analysis. If it is important to you at which exact time during the day a transaction takes place; you could make a dimension table that has a row for every second. By adding time, you can answer questions such as "Do we sell more bread in the morning or during the afternoon?" For now, we will assume that we have a star schema where the date defines the grain and an exact time is not necessary for our intended analysis.

Figure 7.7 shows an example date dimension:

Date	DateKey	Day	DayName	Month number	Month	MonthKey	Year	Fiscal Month	Fiscal Year	day Index	month index	year index	Day number
22-12-2020 00:00:0(	20201222	22	Tue	12	December, CY2020	202012	2020	6	2021	-171	-6	-1	357
23-12-2020 00:00:0(	20201223	23	Wed	12	December, CY2020	202012	2020	6	2021	-170	-6	-1	358
24-12-2020 00:00:0(	20201224	24	Thu	12	December, CY2020	202012	2020	6	2021	-169	-6	-1	359
25-12-2020 00:00:0(	20201225	25	Fri	12	December, CY2020	202012	2020	6	2021	-168	-6	-1	360
26-12-2020 00:00:0(	20201226	26	Sat	12	December, CY2020	202012	2020	6	2021	-167	-6	-1	361
27-12-2020 00:00:0(	20201227	27	Sun	12	December, CY2020	202012	2020	6	2021	-166	-6	-1	362
28-12-2020 00:00:0(	20201228	28	Mon	12	December, CY2020	202012	2020	6	2021	-165	-6	-1	363
29-12-2020 00:00:0(	20201229	29	Tue	12	December, CY2020	202012	2020	6	2021	-164	-6	-1	364
30-12-2020 00:00:0(	20201230	30	Wed	12	December, CY2020	202012	2020	6	2021	-163	-6	-1	365
31-12-2020 00:00:0(	20201231	31	Thu	12	December, CY2020	202012	2020	6	2021	-162	-6	-1	366
1-1-2021 00:00:00	20210101	1	Fri	1	January, CY2021	202101	2021	7	2021	-161	-5	0	1
2-1-2021 00:00:00	20210102	2	Sat	1	January, CY2021	202101	2021	7	2021	-160	-5	0	2
3-1-2021 00:00:00	20210103	3	Sun	1	January, CY2021	202101	2021	7	2021	-159	-5	0	3
4-1-2021 00:00:00	20210104	4	Mon	1	January, CY2021	202101	2021	7	2021	-158	-5	0	4
5-1-2021 00:00:00	20210105	5	Tue	1	January, CY2021	202101	2021	7	2021	-157	-5	0	5
6-1-2021 00:00:00	20210106	6	Wed	1	January, CY2021	202101	2021	7	2021	-156	-5	0	6

Figure 7.7 – Example date dimension

As you can see in *Figure 7.7*, the date dimension has a row for each date on the calendar. The first column stores the actual date as a datetime data type. All other columns are redundant. They can all be calculated from the first column. So, why store them in a date dimension? Why not create a separate table altogether and just use the `OrderDate` column, which you can store in the fact table? Having a separate table has a couple of advantages. These advantages are as follows:

- Creating reports using drag and drop
- Working with fiscal years and other centrally defined business rules
- Dealing with missing dates
- Adding extra functionality
- Considering performance

Let's have a look at each of these advantages in more detail.

Creating reports using drag and drop

As we already said, all columns in a date dimension except the date column itself can be calculated from the date. Without a date table, the person using the data needs to be able to write the correct expressions for these calculations. With self-service BI, we want as many people as possible to be able to create reports and do the analysis themselves. That includes users who do not have the technical skills needed to calculate, for example, a day name in a specific language.

When you need to create a sales-by-year report, just dragging a `SalesAmount` column and a `Year` column to the report is by far the easiest way. This is, of course, only possible when a `Year` column already exists. If there are also reports that show sales by month, sales by quarter, or sales by trimester, we make sure we have columns for month, quarter, and semester. The point is, at the time of actually working with the data, everything that you need is readily available to use. A simple drag and drop will do the trick. Quick and simple.

Working with fiscal years and other centrally defined business rules

It was said earlier that you can calculate all columns in a date dimension from the date column itself. But you might need some understanding of the business to be able to calculate everything. Microsoft, for instance, has a fiscal year starting on July 1. That means that July has a month number equal to 1 and the fiscal year belonging to the calendar date July 1, 2021 is 2022.

In the example in *Figure 7.7*, you can see both **Fiscal Month** and **Fiscal Year** implemented as just described. With these columns available, reporting on fiscal years becomes as straightforward as reporting on calendar years. But to calculate them, you need the technical skills and the business rule that defines fiscal years.

You could calculate the fiscal year at the report level if you have the technical skills. But if you have 10 reports that use fiscal years, you potentially have to write the logic 10 times. You also add the risk of implementing a different business rule in one of those reports, which now makes your total reporting environment inconsistent. A central definition that is defined once and can be reused by as many reports or users as needed is the better solution.

There are other business rules that you might want to define centrally to be used on multiple reports and in multiple analyses. A famous example would be week numbers. Multiple definitions in the world define which week is week 1. From an analytics perspective, it doesn't matter which definition you use. However, a centrally defined definition that is used by all reports means that you get consistency over multiple reports.

Other business rules might, for instance, be the definition of holidays. You may want to analyze whether sales are higher or lower on holidays compared to working days. But what is a holiday? Each country has different holidays. A company could define a list of dates that they consider to be holidays. You can add a column that you, for instance, call `IsHoliday`. For each holiday on the list, you enter the value `True`, or otherwise `False`. Your analysis will now be straightforward using this column.

Dealing with missing dates

Suppose you create a sales-by-month report for a small business. You calculate the month from the order date and you add the individual sales amounts of each order. You might end up with a report with only 11 months on it. If the business was closed during the summer holidays and did not have a single order in the month of August, you will have 11 months only. August doesn't exist in the database at all and you cannot report on data you don't have. But not having sales in August is information even if you don't have data for August. Using a date dimension ensures that you have data for all the dates on the calendar, as well as for dates (or entire months) without rows in the fact table.

An extra argument might be the tool you use to create reports. Power BI and Azure Analysis Services have DAX functions that you can use to calculate time intelligence measures. Most of these functions only work correctly when you have a date dimension. They need a single row for each date on the calendar to work.

Adding extra functionality

If you look at *Figure 7.7*, you can see three different columns for the month. The column called `Month` holds values such as `January, CY2021`. Creating values such as this accomplishes two things:

- The first is that you now have a different label for January 2020 than for January 2021. Suppose you create a report in Power BI by dragging `SalesAmount` to the report and dragging `Month` to the report. With just the `January` value, Power BI will create a bar chart with 12 bars. Sales for January 2020 and sales for January 2021 will be added together and shown in the bar for January. Adding the year to the label means that both months are now actually different and Power BI will treat them as such. "Just drag and drop" will now be sufficient. Otherwise, you will have to do more work in creating your Power BI report. That takes extra time and asks additional skills of the Power BI user.

- The second advantage of the labels we created is that they make the report clearer and with that, the information easier to interpret. Suppose, to use another example, you have a bar chart with four bars. Each bar has a value on the axis – 1, 2, 3, and 4, respectively. What is this chart telling you? Now suppose the values on the axis show labels such as `Q1, 2021` instead of just `1`. It is now immediately clear what the bar chart is about. The labels in a Power BI report are the values in the column that you use for the axis of a visualization. So, make readable, clear labels in your dimension tables.

As another example, you see a couple of index columns in the table in *Figure 7.7*. Let's say you need to create a report in Power BI that shows sales from the last 5 years. The column year index is equal to 0 for each day in the current year. It has a value of -1 for each date in the last year and it has the value of -5 for each day in the year that is currently 5 years in the past. When you recalculate your date dimension every night, these values change over time. But from a Power BI perspective, you can now simply add a static filter specifying you only want dates where the year index is greater than or equal to -5. And that brings us back to making analytics as easy as possible.

A date dimension provides a central location for date-related business rules. It increases the ease of use of the data and the consistency over multiple use cases. Another factor that deserves a little attention is performance.

Considering performance

In *Figure 7.7*, you can see a **Date** column and a **DateKey** column. The **DateKey** column holds a smart key. The value is of the `int` data type. It is just a number. At the same time, it is not a random, meaningless number. It is not an ordinary surrogate key. You can recognize the date in the number in the format of `yyyyMMdd`. The first four digits correspond to the year, digits five and six correspond to the month number, and the last two digits correspond to the day. It used to be best practice to create a key like this, and for some systems, it still is. The underlying idea is that an `int` data type is the most efficient data type to work with. Considering there will probably be a lot of joins between the fact table and the date dimension, these keys are used a lot.

The data preview of *Figure 7.7* suggests that the `Date` column is of the `datetime` or `datetime2` data type. Both these data types are bigger in the number of bytes they use than the `int` data type. A better choice for the data type would be to use `date` because the example shows a date dimension without time. The `date` data type is only three bytes in size, whereas `int` is four bytes in size. When modeling for SQL Server or Power BI, you do not need a special key column of the `int` data type for performance. Using a `Date` column with the `date` data type is better.

Next to the `DateKey` column, the example date dimension of *Figure 7.7* also has a **MonthKey** column. This column is of the `int` data type again. Each month gets an integer value corresponding to the month. This column can help in sorting months correctly. The **Month** column will sort alphabetically, making April the first month of the year. Using **MonthKey** to sort by will provide the correct sorting order.

The important thing that we learned in this section is that you always need a date dimension. The columns you create depend on how you intend to use the date in your intended analytics. Now, let's look at how historical data can play a role in analytics and how to implement historical data in a dimensional model.

Slowly changing dimensions

The first section of this chapter stated that the lack of historical data in source systems is one of the problems you face when performing analytics on operational systems directly. Kimball named this **Slowly Changing Dimensions (SCD)** and has a couple of different solutions on how to store historical data.

The term "slowly" refers to the fact that most dimension attributes only change every now and then. An average customer will move maybe every couple of years, but not 10 times per day. Their address changes "slowly." In an online retailer, a product price might vary from hour to hour or maybe even by the minute. We are not speaking of a "slowly" changing dimension anymore and you should consider the impact of the solution you choose to store all prices.

The "changing dimension" part might lead to some confusion. SCD refers to attributes of a dimension for which values may change over time. Every individual attribute in a dimension table might change. You need to consider for each attribute individually how to handle changing values in your dimension table. The meaning of the attribute and, more importantly, how you use the attribute, should lead when choosing a solution.

The theory of dimensional modeling has three standard solutions on how to deal with changing attribute values. These solutions are as follows:

- SCD type 1
- SCD type 2
- SCD type 3

Let's look at all three solutions in turn.

Implementing SCD type 1

Suppose you design a `dimSalesPerson` dimension that has a `PhoneNumber` attribute. You will most likely not create an analysis that shows sales by phone number. You may, however, create a report to show sales by salesperson. When a user of the report hovers their mouse over the last name of the salesperson, a popup might show the salesperson's email address and phone number. This enables the report user to reach out to the salesperson directly. For this, you want the phone number to always be the salesperson's current phone number. Whenever a salesperson gets a new phone number, you want the old number to be overwritten by the new number. Doing so is called SCD type 1.

> **Note**
> With SCD type 1, the old value of an attribute is overwritten by the new value. Values are not tracked historically.

When loading new data into a dimensional model, you will have to compare the values in the star schema to the new values (coming from the operational systems). If you find an existing row in the dimension table describing a salesperson, but the value in the star schema is different to the value coming from the operational system, you have to change the existing row in the dimension table. The ETL process is responsible for this. You will learn about ETL processes in *Chapter 8, Provisioning and Implementing an Azure Synapse SQL Pool*, and *Chapter 11, Implementing ETL Using Azure Data Factory*.

You could choose to keep the old phone number in the dimension table. Keeping the original value instead of the most current value is called SCD type 0. This type is used very rarely.

Let's continue to look at what to do when you do want to track changes.

Implementing SCD type 2

Suppose the `dimSalesPerson` dimension also has a `SalesRegion` attribute. Each salesperson has a sales region assigned to them. Sales regions with a lot of customers may have multiple salespersons working in the region. Also suppose that you create a report that shows sales per sales region. What would happen to this report if you changed the value of `SalesRegion` of an existing salesperson with existing facts in the fact table?

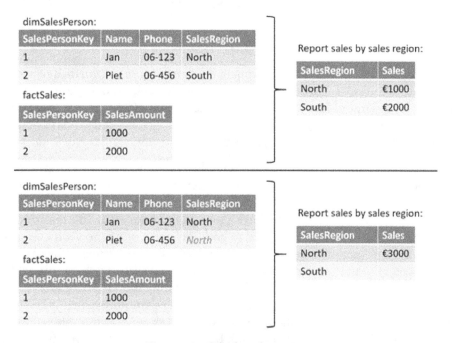

Figure 7.8 – Sales by sales region

Look at *Figure 7.8*. Above the line on the left-hand side, you see a **dimSalesPerson** table with two salespersons. Jan works in the sales region called **North** and Piet works in **South**. The **factSales** table shows that Jan has 1,000 euros in total sales amount and Piet has 2,000 euros. The sales-by-sales region report then shows 1,000 euros for the **North** sales region and 2,000 euros for the **South** sales region.

Under the line in *Figure 7.8*, the **dimSalesPerson** table has a different value for **SalesRegion** for salesperson Piet. Piet moved from **South** to **North** on the first of February. But simply changing the value for **SalesRegion** messes up the sales-by-sales region report. The 2,000 euros that were already in the database still belong to the **South** sales region. New orders coming in for Piet have to be registered to **North**, but old orders are still orders from the **South** region. If you want to create the sales-by-sales region report correctly, you cannot simply overwrite the **SalesRegion** value.

In these kinds of scenarios, you use SCD type 2. With SCD type 2, you create a new row whenever an attribute value changes. Have a look at *Figure 7.9*, which shows the same dimension, **dimSalesPerson**, but now with an SCD type 2 implementation:

SalesPersonKey	SourceKey	Name	Phone	SalesRegion	StartDate	EndDate
1	1	Jan	06-123	North	1-Jan-2000	31-Dec-9999
2	2	Piet	06-456	South	1-Jan-2000	15-Feb-2021
3	2	Piet	06-456	North	16-Feb-2021	31-Dec-9999

Figure 7.9 – SCD type 2

When salesperson Piet changed from working in the **South** region to **North**, a new row for salesperson Piet was created. We added two columns to the table, **StartDate** and **EndDate**. We end-dated the old row of Piet with the end date of **15-Feb-2021**. Up until February 15, 2021, Piet worked in the **South** sales region. A new row for Piet was added with a **StartDate** value of **16-Feb-2021**. From February 15, 2021 onward, Piet works in the **North** sales region.

With a new row, you also get a new primary key value. From now on, every new fact row that enters the database for Piet will have a foreign key value of 3. That relates new facts to Piet but also to North. All existing fact rows for Piet, rows that already exist in the database prior to Piet moving to North have a foreign key value of 2. This doesn't change. That means those older facts remain related to both Piet and South. You can now safely create reports to show both sales by salesperson and sales by sales region. When you use Power BI, for example, and you drag **Region** to the report, Power BI will create a table visual with the values **North** and **South**. When you next drag the **SalesAmount** column to the table visual on the report, Power BI will sum all **SalesAmount** values linked to **SalesPersonKey** 1 and 3 and show them beside **North**. Both rows have a value of **North** for the region. Likewise, when you do not drag **Region** to the report but you use **Name** instead, all **SalesAmount** values linked to **SalesPersonKey** 2 and 3 will be added together because these rows have the same value for **Name**.

Notice that the EndDate column has a value of 31-dec-9999 for rows that store the current values for all attributes, the current row. We could have left this value empty because technically, there is no end date. However, using 31-dec-9999 is easier than using NULL when writing queries and especially when using date ranges in queries.

Detecting change, end-dating rows, and creating new rows is again the responsibility of the ETL process.

> **Note**
>
> With SCD type 2, each change leads to end-dating the old version of the row and creating a new version of the same row.

One thing to be aware of when implementing SCD type 2 changes is that you get multiple rows for, in this case, single salespersons. Simply counting the number of rows does not give you an accurate number of how many salespeople there are. Also, when changes become frequent, your table will grow rapidly. That is why this is a good solution for *slowly* changing dimensions and not for *rapidly* changing dimensions.

If you need to keep track of a rapidly changing attribute, you might consider creating a so-called mini dimension. A mini dimension only stores attributes that change frequently instead of all attributes logically belonging to the same dimension table. In the example of product prices that change frequently, you create a "normal" product dimension and a separate one storing only the product price. You now treat price as a regular SCD type 2. Because this table only stores price information, it doesn't hurt so much that you get a lot of rows implementing SCD type 2.

Now that we have discussed how to historically track changes using SCD type 2, let's look at an alternative, SCD type 3.

Implementing SCD type 3

Suppose you categorize customers into the Platinum, Gold, and Silver categories based on what the customers' assigned account manager expects that this customer will purchase in the next 12 months. You want to keep track of how many customers change from one category to another.

You clearly need to keep track of the category historically. You could use an SCD type 2 implementation to keep track of the history. But using SCD type 2 here might lead to complex SQL or complex DAX to get to the requested insight. And we know by now that complexity leads to errors and bad performance. SCD type 3 can help. Have a look at *Figure 7.10*:

CustomerKey	SourceKey	Name	CurrentCategory	PreviousCategory	DateChanged
1	1	Jan	Gold	Platinum	18-1-2020
2	2	Piet	Silver	*NULL*	1-1-2000

Figure 7.10 – SCD type 3

SCD type 3 creates two columns for the **Category** attribute, **CurrentCategory** and **PreviousCategory**. Whenever a customer's category is changed, you update the existing row describing that customer. The new category is written in the **CurrentCategory** column and whatever value was stored in **CurrentCategory** is copied to the **PreviousCategory** column. You register the date the change took place in a column called **DateChanged**.

Using this approach, the requested report is just applying the proper filters. All information needed is in a single row; you need not match a row to another row that has the previous value stored.

> **Note**
> With SCD type 3, you create a current and a previous column storing both the current and the old value of an attribute in the same row.

The most notable limitation of SCD type 3 is that you only store one version back. What if a customer changes categories again? You will lose the category they belonged to before they had the category currently stored in the **PreviousCategory** column.

Because of this limitation, SCD type 3 is not often used. However, especially in Power BI, it may be worth the trouble to solve the issue as described here in Power Query instead of in DAX. DAX expressions for calculated measures are calculated during report creation. These calculations may take some time to complete, which could lead to poor performance of the report. Power Query logic is executed during dataset refresh. Your users might not even be aware of when that takes place. You can also oftentimes schedule the refresh to take place during the night. Users will not notice how long the refresh takes.

In the original theory of Ralph Kimball, only types 1, 2, and 3 were discussed. Nowadays, a lot of alternatives have emerged. Most of them are smart combinations of the three basic types of SCD.

Before we move on to discussing facts in more detail, there are two special kinds of dimensions that deserve some explanation. We'll start by looking at the junk dimension.

Junk dimension

Some dimensions can be really small. Think, for instance, of a `dimPaymentMethod` dimension, where a webshop only allows payment by credit card or iDeal. The dimension will have one column (apart from the key), `PaymentMethod`, and two rows, one for credit card and one for iDeal.

This is no problem in itself. But suppose you have a similar small dimension describing, for instance, the shipping. Some orders are eligible for free shipping, while others are not. And now suppose you have 10 dimensions like this.

When you create a dimension, in each case you end up with a lot of foreign keys in the fact table. This makes the fact table inefficient and the star schema less intuitive. You could instead create what is called a junk dimension. This is a dimension where you combine unrelated small dimensions into a single dimension table. You create rows for each combination of values. *Figure 7.11* shows an example based on the payment type and shipping method mentioned previously:

dimJunk
JunkKey
PaymentMethod
Shipping

JunkKey	PaymentMethod	Shipping
1	iDeal	Free
2	iDeal	Paid
3	Creditcard	Free
4	Creditcard	Paid

Figure 7.11 – Junk dimension

The junk dimension of *Figure 7.11* has only four rows. There are two rows for payment and two rows for shipping. That makes four possible combinations of payment method and shipping. If both underlying dimensions have 1,000 options, the junk dimension would have 1 million rows. It only makes sense to create a junk dimension when you have a couple of attributes that do not belong to any dimension and only have a few distinct values.

When you have a single column dimension with a lot of values, a degenerate dimension might make more sense.

Degenerate dimension

Consider, for example, an `OrderNumber` attribute. You could do something such as sales by order or average sales amount per order. That makes `OrderNumber` a dimension attribute. However, a `dimOrder` dimension table would have a single `OrderNumber` column. It does not make sense to create a table for this single column. Adding it to a junk dimension also doesn't make sense because there will be a lot of different order numbers. The solution is to add an `OrderNumber` column to the fact table. We call that a degenerate dimension.

Be careful with degenerate dimensions in Power BI and Azure Analysis Services. DAX filtering will work differently when the used dimension attribute is part of the fact table compared to when it is in an actual separate dimension table. Try to avoid degenerate dimensions in these two products. When implementing a star schema in Azure SQL Database or Azure Synapse Analytics, it can be helpful to implement degenerate dimensions.

Dimension tables are really important. But it is now time to explore fact tables in more detail.

Designing fact tables

Dimension tables may be important, but the facts are what really matter. So, before going into fact tables, let's look at the facts first. We already learned about facts in the dimensional modeling section, but we need to look at facts in a bit more detail. When we do, we can distinguish three types of facts:

- Additive facts
- Semi-additive facts
- Non-additive facts

We will discuss them in turn.

Understanding additive facts

Additive facts are numerical facts that you can add together to create facts at a higher aggregation level. Almost all reports use aggregated facts. For instance, a report showing sales by month is an aggregated report. All individual sales transactions of the same month are put together to form one new row. The sales are calculated by adding all the sales amounts together to form the month's sales. The overall sales are then calculated by adding all the sales amounts of all the months together.

Next to sales by month, you could create sales by product, sales by customer, or sales by country, to name just a few possibilities. Again, the sales amount is calculated by adding (summing) all the individual amounts together to make one value.

A fact that can be aggregated by adding individual detail-level values is called an additive fact. Most BI tools use the SUM function as the default function to aggregate values. When you create a report in Power BI, by dragging `SalesAmount` and `OrderDate` onto the report, Power BI will automatically create a date hierarchy and show `SalesAmount` by year using the SUM DAX function to calculate the yearly sales amounts shown. Sales is a common example of an additive fact.

> **Note**
>
> An additive fact uses the SUM function to aggregate detail-level values.

Understanding semi-additive facts

Not all facts can be aggregated using the SUM function. Think, for instance, about a fact such as inventory or your balance in your bank account. Suppose you have 500 dollars in your bank account. Your friend has 600 dollars in their account. Together, you have 1,100 dollars. Or, as another example, you have 500 dollars in your checking account and you have 700 dollars in your savings account. Again, we can add the two values together, and you correctly say that you have a total of 1,200 dollars.

But what happens when we analyze your balance using the date dimension? You have 500 dollars at the end of January, you have 600 dollars at the end of February, and 500 again at the end of March. Do you now have 1,600 dollars at the end of quarter 1? This is what Power BI will show you unless you overrule its default behavior. Bank balance is an example of a semi-additive fact. You can use the SUM function when aggregating unless you aggregate over a period. In that case, the aggregate should be the last value. The balance at the end of quarter 1 is the same as the balance at the end of March, not the sum of January, February, and March.

> **Note**
>
> A semi-additive fact can be calculated using the SUM function for all dimensions except when analyzed using the date (or time) dimension.

Understanding non-additive facts

Non-additive facts can never be calculated using the SUM function. A good example is a percentage. Suppose you have a margin of 5% over January and a margin of 6% over February. That does not add up to a margin of 11% over the first 2 months. Taking the average of 5.5% also doesn't make any sense. You probably need a weighted average.

Analyzing your margins by, for example, product category doesn't make any difference. If you have a margin of 5% for product category A and 6% for product category B, you do not have an overall margin of 11%.

> **Note**
>
> Non-additive facts cannot be computed using the SUM function.

Power BI will automatically assume that each numerical column is a fact and Power BI will use the SUM function to calculate aggregates. That means that you want to be careful with creating semi- and non-additive measures. You might want to create measures such as SalesAmount and CostAmount and leave it to Power BI to calculate the margin you made based on these two facts. Both mentioned facts are additive and have a low risk of being used wrongly in Power BI, and Power BI needs a DAX expression to calculate semi- and non-additive measures anyway.

Now that we know that there are different types of facts, let's look at different types of fact tables. As with facts, there are three different types of fact tables:

- Transactional fact tables
- Periodic snapshot fact tables
- Accumulating snapshot fact tables

Understanding transactional fact tables

In a transactional fact table, each row corresponds to an event or transaction that took place. A row is the one-time registration of that event. Take again, for example, a supermarket. Each time a product is scanned by a cash register, a row is entered into the database. You register each individual event of a product being scanned.

An important characteristic of transactions as referred to here is that they are one-time events that generate immutable data. You may store, for instance, the price of the product and the quantity in which the product is bought, together with a timestamp and a cashier, and those "facts" will forever stay the same. A new scan is a new transaction, which is a new row in the fact table. Once scanned, the quantity, price, timestamp, and cashier do not change anymore.

The type of questions you can answer using a transactional fact table are questions such as "How many?", which are answered by counting or summing values. Think about questions such as "How much bread is sold?" or "How much bread did we sell in June 2020 in Amsterdam?" as well as questions such as "What was the total revenue of our London-based store in 2020?" Comparing the answers to previous periods to calculate something such as growth is, of course, possible as well.

Questions such as these are critical in a lot of environments. That makes transactional fact tables widely used.

Understanding periodic snapshot fact tables

A periodic snapshot fact table is not the registration of an event, but the registration of a situation at a specific point in time. Inventory can be a good example. We can store a row each time a product is purchased or sold. You can then calculate the current inventory of a product by adding all the quantities you purchased and subtracting all the quantities you sold. The balance in your bank account can be calculated in a similar way by adding and subtracting all the amounts of all individual transactions. In both cases, the queries you need to do the calculations will be heavy and long-running. The transactional approach we just described may not be the best solution. A periodic snapshot fact table can help.

A snapshot is a point-in-time registration of the situation at that specific time. *Figure 7.12* shows a simple example:

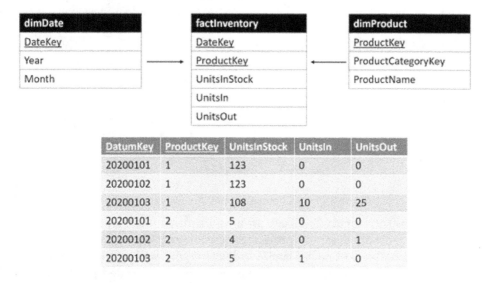

Figure 7.12 – Periodic snapshot fact table

The table in *Figure 7.12* has a row for each product for each day. You can see that we had 123 units in stock of product 1 on January 1, 2020. Because nothing happened, the situation was the same on January 2. However, on January 3, we sold 25 units of the product and purchased 10 units. The new inventory at the end of the day was 108 units. Each row shows the number of units that left and entered our inventory and the closing situation of the day for a specific product.

This type of fact table makes questions such as "How many products did we have in stock at the end of January?" easy. Showing the fluctuation of inventory over time is also easy using this type of fact table.

Understanding accumulating snapshot fact tables

The third and last type of fact table is the so-called accumulating snapshot fact table. Suppose you have a sales process that starts with an opportunity. You register that a potential customer showed interest in a product, but they weren't ready to buy yet. After a while, that potential customer might ask you for a quote. You making a quote in this scenario is an event, but the quote is not an entity. It is merely the already-registered opportunity moving into the next phase of the sales process. The quote can later become an order, which in turn becomes a delivery, which becomes an invoice, which becomes a payment. Even though a payment is not an invoice and an invoice is not a delivery, it is all the same fact, just at a different stage in the sales pipeline.

You could create a transactional fact table with all opportunities, and then a separate one for all quotes, and a third one for all orders, and so on. Or you could add each event as a separate row in the same fact table. In both cases, the query that will give you insight into lead times will be difficult. You might want to answer questions such as "What is the average duration between a quote and an order?" or "How many quotes are converted into orders?" easily. Accumulating snapshot fact tables are designed for these types of scenarios. *Figure 7.13* shows an example:

Order Key	Customer Key	Product Key	Opportunity Date	Quote Date	Order Date	Delivery Date	Quantity
1	1	1	1-Jan-2020	12-Jan-2020	13-Jan-2020	15-Feb-2020	5
2	2	2	2-Jan-2020	15-Jan-2020	NULL	NULL	6

Figure 7.13 – Accumulating snapshot fact table

An opportunity arose on January 1, 2020. Customer 1 showed interest in product 1. On January 12, we created a quote for this product for customer 1. They signed the quote, making it into an actual order one day later. We delivered the product to this customer on February 15. Another opportunity was created on January 2. We created a quote for this opportunity on January 15. That is the last-known status of this quote. It could turn into an order at a later date, or it may never become an order, in which case the **OrderDate** column will always keep the value **NULL**.

The fact table shown in *Figure 7.13* represents a sales pipeline where a new stage is registered by entering the date at which a new stage is entered. We can easily create facts such as duration between `QuoteDate` and `OrderDate` in days and use that fact on reports with an `AVERAGE` function.

Having multiple dates in a fact dimension leads to multiple relationships between the fact table and the date dimension. This is called the roleplaying dimension.

Understanding the roleplaying dimension

The fact table shown in *Figure 7.13* has four date columns. Each column is a foreign key to the `dimDate` dimension table. If your date dimension uses smart keys, you need to use smart keys in the fact table as well. Otherwise, real dates, such as in *Figure 7.13*, will suffice. With four foreign keys, you have four relationships between the fact table and the dimension table. This is called a roleplaying dimension. The `dimDate` dimension plays the role of `OrderDate`, but also the role of `DeliveryDate`. This means you use the same `dimDate` dimension to analyze the `Quantity` fact by `OrderDate` as when you analyze the `Quantity` fact by `DeliveryDate`, or one of the other dates. In SQL queries, you need to write a different `JOIN` clause; in Power BI, you can use the `USERELATIONSHIP` DAX function.

You know by now the three types of fact tables we distinguish between. But sometimes, your analysis needs to focus on what did *not* happen. In all the mentioned fact table types, you store what *does* happen. A **coverage fact table** may help.

Using a coverage fact table

Suppose you have special actions, promotions, for specific products during a specific period. To analyze the impact of your special actions, you create a dimension table called `dimPromotion` with all the actions. For each order, you register which promotion applies. You can create a "dummy" promotion called "not applicable" for all sales where no special action was applied.

This setup can get you into trouble if you do not sell the product in the period when the promotion was active. Suppose `Action1` applies to `ProductA` and `ProductB` during week 1 of 2021. You do sell `ProductB` during that period, so a sales row is added to the fact table and the row is linked to the action in the `dimPromotion` table. But `ProductA` is not sold during week 1. This cannot be found because there is no row.

In this situation, you need to create an extra fact table with two columns, that is, two keys: `PromotionKey` and `ProductKey`. These columns are foreign keys to the promotion dimension and the product dimension, respectively. This is where you can keep track of which special action is applicable for which products, independent of actual sales. This special fact table is sometimes called a coverage fact table, covering all the combinations between products and actions.

Figure 7.14 shows an example of a coverage fact table:

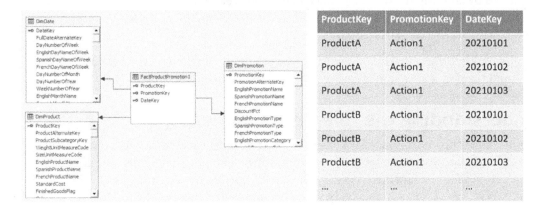

ProductKey	PromotionKey	DateKey
ProductA	Action1	20210101
ProductA	Action1	20210102
ProductA	Action1	20210103
ProductB	Action1	20210101
ProductB	Action1	20210102
ProductB	Action1	20210103
...	...	...

Figure 7.14 – Coverage fact table

There are a couple of design alternatives when it comes to coverage fact tables. In *Figure 7.14*, we added a date column to keep track of the specific dates that the promotion was applicable. You could also create a single row per product and promotion combination and add a validity period to the **dimPromotion** table.

The fact table as described here is also an example of a **factless fact table**. The fact table doesn't store any real facts. The facts are in the combinations that you create.

Now that we have learned all about facts and dimensions, it is time to have a look at the bigger picture. What does the entire data warehouse look like, or do we want separate data marts?

Using a Kimball data warehouse versus data marts

The starting point of creating a star schema is choosing a process to model. One star model describes one process. A business is always more than a single process. In the theory of Ralph Kimball, a data warehouse is the collection of all star schemas that together describe the entire organization.

There will always be overlap between the individual star schemas you create to model the individual processes. The processes are not completely independent of each other. The sales department sells products that the purchasing department buys. They work with the same products. So, the star schema describing the sales process will have the dimProduct and dimDate dimensions in common with the star for purchasing. They will have different dimensions as well. The star schema for purchasing might have a dimension for suppliers, whereas the sales start schema probably has a dimension for customers.

When you design the `dimProduct` dimension table for the sales process and you design it again, completely independent from the first design, for the purchasing process, you probably end up with two different tables. Even though both sales and purchasing work with the same products, they probably need to know different things about the products. Purchasing might need to know what the delivery time of a specific product is and what the minimum stock level is for each product. Sales doesn't care about the delivery times of suppliers, but might need to know how much VAT to include in the quote they make for a customer. The product dimension appears differently for the sales process than for the purchasing process.

Whether the situation described is a problem or not depends on the situation. If you need to implement a Kimball-oriented data warehouse, you need to be careful. A Kimball data warehouse is where you combine the star schemas of the individual processes into a single big database. Dimensions such as `dimProduct`, as described, need to be reused in all processes that have something to do with the product. You want to create one product dimension, `dimProduct`, and use that single table in both the sales and the purchasing star schemas. For this to work, you need to create a so-called **conformed dimension**. A conformed dimension is a dimension that you design in the most generic way possible in order to make it useful for all (potential) star schemas. The `dimDate` dimension should have columns for `DeliveryTime` and `MinimalStockLevel` to accommodate purchasing, but also a `VATPercentage` column to accommodate sales.

In modern business intelligence architectures, you might not want to create a completely dimensionally modeled data warehouse. In *Chapter 10*, *Designing and Implementing a Data Lake Using Azure Storage*, we will explain the modern data warehouse architecture with a **data lake** and the use of **data marts**.

A data mart is a subset of all the data you have specifically optimized for a use case. In a modern data warehouse architecture, you store all your historical data in a data lake. When you need to create a sales dashboard and reports, for instance, you take the relevant data for this specific purpose out of the data lake and put it in a data mart. Since a star schema is the most optimized design for reporting and analytics, you use dimensional modeling to design the data mart. Because this data mart is created specifically to provide information to the sales department, you don't have to consider how purchasing looks at products. You can design the product dimension completely optimized for sales.

If, at a later date, purchasing wants the dashboard and reports as well, you design a star schema for them. This star schema can then get its own product dimension. It is completely independent of any other data mart that you created earlier. This does mean, however, that you now have two product dimensions that probably overlap with each other for the most part. This is inefficient. However, the flexibility of having independent data marts and the usability of an optimized dimension table outweigh that inefficiency.

You might get into a situation where you need to do cross-process analysis. If a supermarket buys a lot of cans of soup and they build a wall of soup cans that a customer can hardly avoid, the sales of soup go up. So, you may want to analyze how sales and purchasing affect each other. Supply might create demand. In such a case, you create a data mart that might have fact tables for both purchasing and sales. You now want them to both use the same product dimension. It needs to be confirmed again.

Something similar as with data marts is the case for Power BI. Power BI works best when the tables in Power BI form a star schema. When you start creating a report in Power BI, you already have a specific goal in mind. You can use Power Query to transform the source data into a star schema if the source isn't already modeled dimensionally. This star schema is, of course, optimized and geared completely to the intended goal.

Sometimes you might use Power BI to create a dataset that you publish to the cloud. You let self-service BI users create their own reports using your published dataset. This dataset acts as a data mart to your self-service BI users. Because you might not know exactly what your users are going to do with the data, you want to keep the dimension generic. The more attributes they have, the more different types of analysis they can perform.

Summary

Kimball invented dimensional modeling to create an optimal user experience when using a relational database as a source for reporting and analytics. The design is optimized to read data. The design is also optimized for large datasets.

A star schema should be intuitive. It should be easy to use. It improves the productivity of report developers. It also decreases the number of errors made. This makes star schemas well suited for self-service BI where people with potentially few technical skills but lots of process knowledge work with the data.

In this chapter, you learned the skills to design a star schema database model. You learned to design both dimension and fact tables.

In the next chapter, you will implement a star schema design using Azure Synapse Analytics.

Exercise

Figure 7.15 shows the ERD of Northwind that you saw in *Chapter 2, Entity Analysis*. Northwind is a company that purchases delicacies (products) from suppliers all over the world. The products have been categorized into categories. Products are sold to customers by Northwind employees. Shippers make sure that customers receive the products they buy:

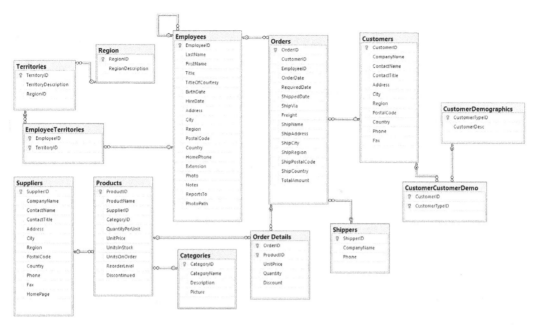

Figure 7.15 – Northwind ERD

Because management feels they lack control because they lack information, you are asked to create management reports based on a star schema database. Following an initial analysis, the following demands have been formulated:

- The sales department needs reports showing sales by customer, by salesperson, by product, by supplier, and by country:

  a. All reports should be based on fiscal years starting on July 1.

  b. An analysis needs to be broken down from years into quarters and months. When an analysis is performed based on the calendar rather than on fiscal years, trimesters are often used.

  c. A sales by employee analysis uses age category and the number of years in service. There are two age categories: younger than 50, or 50 years or older. Years in service are categorized into bins of 10 years.

  d. Sales is contributed to the country the paying customer lives in, not the country where the products are shipped to.

- Customer service needs to get insight into the time between ordering and the delivery of the ordered products.

- The purchasing department needs to be able to track inventory over time.

- In the near future, Northwind will start using Power BI. The data warehouse should therefore contain as much information as possible.

Exercises

- Which fact tables should you create?
- What grain should your fact tables have?
- What dimensions do you need?
- Design all the facts and dimensions.
- Create an ERD.

8

Provisioning and Implementing an Azure Synapse SQL Pool

Azure Synapse Analytics is an integrated environment used to create data analytics platforms at scale. The SQL pools of Synapse Analytics are an ideal platform to implement star schema analytical databases at scale.

In this chapter, you will learn about Synapse Analytics. You will implement a star schema database using Synapse SQL pools. The following topics will be explained:

- Overview of Synapse Analytics
- Provisioning a Synapse Analytics workspace
- Creating a Synapse SQL pool
- Implementing tables in Synapse SQL pools
- Understanding workload management

- Using PolyBase to load data
- Connecting to and using a dedicated SQL pool

Overview of Synapse Analytics

Synapse Analytics is a platform comprising multiple components that can all work together to provide an end-to-end analytical solution. As explained in *Chapter 7, Dimensional Modeling*, performing analytics on operational databases comes with challenges and limitations. In a modern environment, you gather data from various source systems in a data lake. The data in the data lake might need validation and curation. The data can then serve as a source for a dimensionally modeled data mart that acts as a serving layer to enable users to actually use the data.

This process involves moving data from one layer to the next, processing the data, transforming the data, and serving the data to users and applications such as Azure Analysis Services and Power BI. Synapse Analytics brings all this together in one suite. It comes with Synapse Analytics Studio, a web-based environment to create and manage the entire solution. *Figure 8.1* shows a schematic of Azure Synapse Analytics:

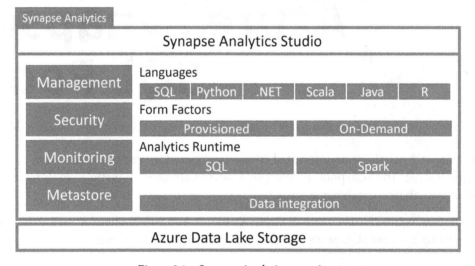

Figure 8.1 – Synapse Analytics overview

The three main components of Azure Synapse Analytics that you see in *Figure 8.1* are as follows:

- SQL pools
- Spark pools
- Data integration

Introducing SQL pools

A Synapse SQL pool is a scale-out implementation of SQL Server. That means that it is a SQL Server database implemented on cluster technology. You can see a high-level architecture of a Synapse SQL pool in *Figure 8.2*:

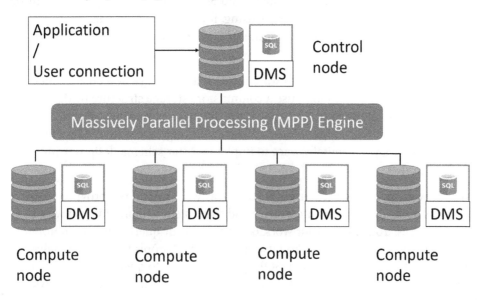

Figure 8.2 – Synapse SQL pool architecture

Figure 8.2 shows a **control node**. This is a SQL Server database that you connect to. You connect in the same way as with an Azure SQL database with normal tools such as **SQL Server Management Studio** (**SSMS**), Azure Data Studio, and Power BI. You use T-SQL to create tables and query the tables.

You also see four compute nodes in *Figure 8.2*. Under the hood, you get 60 SQL databases for every SQL pool you create. These databases are called distributions. All the data you enter in the SQL pool is in reality distributed over these 60 distributions. Queries can be run in parallel where all 60 databases work together on a single query. This is what in the figure is called **Massively Parallel Processing** (**MPP**).

You pay for an Azure Synapse SQL pool by provisioning **Data Warehouse Units** (**DWUs**). The cheapest SQL pool has 100 DWUs. In this case, all 60 distributions are located on the same **compute node**. A compute node is a server hosting the SQL databases that are the distributions. If you provision more DWUs, you will get more compute nodes. With the most expensive SQL pools, you can provision each distribution hosted on its own compute node. You always have 60 distributions. The number of compute nodes you get to spread the distributions over varies between 1 and 60 depending on the number of DWUs you provision.

Synapse has two different types of SQL pools: dedicated pools and serverless. Dedicated SQL pools have dedicated hardware reserved for you. This provides predictable performance and cost, but you also pay for the systems at times you do not use the database. Also, you pay as much per hour during busy times and times where the system is not used that much. One nuance that should be added though: you can pause and resume a dedicated SQL pool, either manually or by using, for instance, Azure automation. When you pause a SQL pool, you keep paying for the storage you are using but you don't pay for compute.

You use dedicated pools to create a data warehouse or a data mart. You create tables and load data into the tables. These tables use a columnstore storage structure by default on top of the distributed architecture. This makes the database extremely well suited for data warehousing workloads with huge volumes of data. Although the system doesn't prescribe a specific table design, you generally want to implement star schema designs using Synapse SQL pools.

Each Synapse workspace has a serverless SQL pool as well. You do not provision DWUs for a serverless pool. You just execute queries and the system will provide you with the necessary resources. You pay for your workload, that is, your queries. If you do not execute any queries, you do not pay. You use serverless SQL pools for **data virtualization** scenarios. In these scenarios, you use T-SQL against Synapse to query data stored in files in your data lake. SQL serverless (until v3) only works on the data stored on the external Data Lake, not on the internal storage of a SQL pool.

Next to SQL pools, Synapse Analytics also has Spark pools. Let's take a look at Spark pools.

Introducing Spark pools

To understand the use cases for Spark pools, consider a modern data warehouse architecture as shown in *Figure 8.3*:

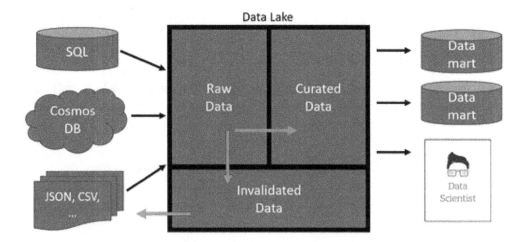

Figure 8.3 – Modern data warehouse architecture

Data is exported from all sorts of systems and copied into the raw section of the data lake. The data inserted in the raw section is almost never directly usable in analytics scenarios. So, in *Figure 8.3*, the data is validated first and invalid rows are moved to the **Invalidated** section of the data lake. Validated rows are now processed and moved to the curated section. Whereas we used to do a lot of data transformations in staging tables in a SQL database, we have now moved those transformation steps to transforming data in files in the data lake.

One operation to perform on the data might be to standardize data. You might get files where dates are stored in the MM-dd-yyyy format whereas other files have dates stored in the dd-MM-yyyy format or even other formats. Making all dates formatted alike simplifies working with the data. Depending on the data and its origin, you might have other operations that you want to perform. You could add another section to the data lake of *Figure 8.3* called **Standardized**.

Curated data is already transformed a bit using business rules. The idea behind a data lake is to store raw data. You add business rules to the data when you know how to use the data. In *Figure 8.3*, the business rules are implemented in the data marts. You might, for instance, implement a star schema data mart for the **Sales** department using a Synapse dedicated SQL pool. This data mart is where all sales-specific business rules are added to the data.

But what if you have generic rules governing your data that always apply? That is, business rules that are specific to the data, not the processes or departments using the data. You do not want to implement the same logic over and over again each time you create a new data mart for a new use case. Applying these generic rules changes the data from standardized data into curated data.

Now back to Spark pools in Synapse Analytics. Depending on the volume of the data that needs to be transformed from raw into curated, you might need a lot of compute power. But that required compute power also depends on how complex the transformations from raw to curated are. Spark is an open source, in-memory big data compute environment that provides the functionality and compute power needed to transform data at scale. Synapse Analytics brings the Spark engine into the same environment as the SQL pool. You use this environment to process the data using Spark and serve the resulting data to consumers using SQL pools.

In *Figure 8.1*, you see the SQL, Python, .NET, Java, Scala, and R languages listed. Whenever you use Spark to do your data processing, you can choose between these languages. Which language you choose depends on the type of transformations you need to do and the experience and skills of the data engineer programming the transformations.

In *Figure 8.3*, data scientists are mentioned as well. A second use case of Spark pools is to leverage the power of languages such as R and Python to do machine learning. Spark itself has extensive libraries for machine learning as well. Using a Spark pool, a data scientist can work with notebooks to do their own data processing and train models using the compute power of Spark. They can use the data stored in the data lake to do their advanced analytics. SQL pools are more geared toward classic descriptive and diagnostic analytics. Spark pools bring predictive and prescriptive analytics into the picture.

With SQL pools and Spark pools, we still haven't seen the entire picture. A third important component of Synapse Analytics is data integration.

Introducing data integration

We just discussed how Spark pools can be used to transform data in your data lake. This is part of the ETL (or ELT) process. **ETL** stands for **Extract, Transform, Load** and ELT switches the order in which you do these individual steps. Roughly, you can say that we extract data from the source systems on the left of *Figure 8.3*, transform the data, and then load the data into a data mart on the right-hand side of *Figure 8.3*. The data marts are the serving layer where the data is ready for usage. ELT is nowadays often chosen over ETL because an ETL pipeline might not be able to handle the incoming data fast enough. Whatever strategy you use, you will need to apply transformations to data and choose the proper tool to do so.

The data integration part of Synapse Analytics brings Azure Data Factory into Synapse Analytics. You will learn more about Azure Data Factory in *Chapter 11*, *Implementing ETL Using Azure Data Factory*.

Another data integration feature of Synapse Analytics is what is called Azure Synapse Link. This feature links Synapse Analytics directly to an operational Cosmos DB. This creates a copy of the data stored in Cosmos DB and stores the copy in a columnar way. You can query this data directly using Synapse SQL pools or Synapse Spark pools. Cosmos DB will keep the columnar copy up to date with the real data in near real time.

The last thing to note about *Figure 8.1* and Synapse Analytics in general is Synapse Analytics Studio. Synapse Analytics Studio is a web-based user interface where all the previously mentioned capabilities come together. You can work with SQL pools, Spark pools, and data integration in a single place. You can also configure a GitHub repository to store all your development efforts safely in a central location.

Now that you have a general understanding of what Synapse Analytics is, let's start using it. This chapter focuses on implementing a star schema in a Synapse Analytics SQL pool. For that, you need a Synapse Analytics workspace. So, that is where we will start.

Provisioning a Synapse Analytics workspace

Let's get straight to the point:

1. Open a browser and go to the Azure portal.

2. Open the portal menu if necessary and click on **Create a resource**.

3. Select the textbox that currently reads **Search the Marketplace** and type Synapse Analytics, and then hit *Enter*.

4. Click, in the list of tiles that fill the screen, on the tile that reads **Azure Synapse Analytics**.

5. Click, on the page that opens, on the **Create** button.

After clicking **Create**, you should see the page shown in *Figure 8.4*. Your screen might look different because Microsoft changes the layout of the Azure portal regularly. However, most options should be available:

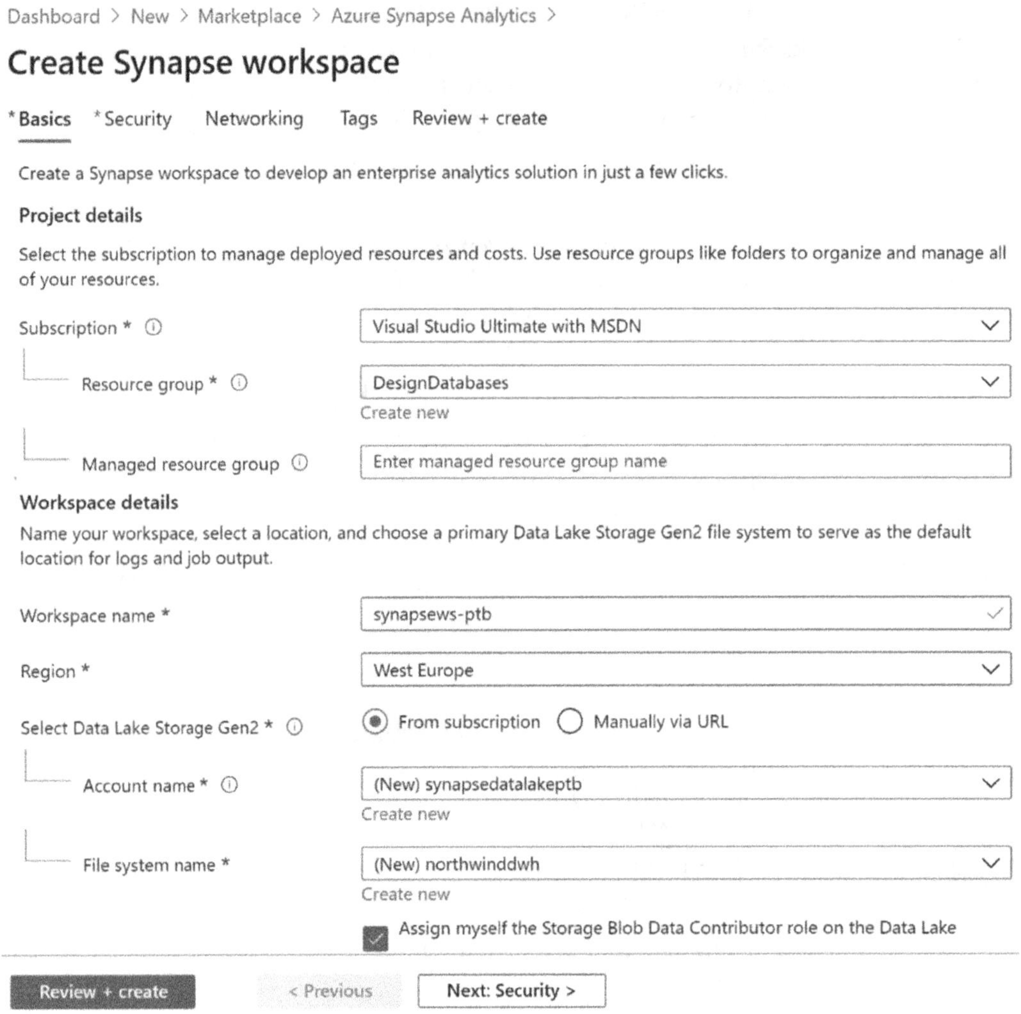

Dashboard > New > Marketplace > Azure Synapse Analytics >

Create Synapse workspace

*Basics   *Security   Networking   Tags   Review + create

Create a Synapse workspace to develop an enterprise analytics solution in just a few clicks.

Project details

Select the subscription to manage deployed resources and costs. Use resource groups like folders to organize and manage all of your resources.

Subscription * ⓘ	Visual Studio Ultimate with MSDN ⌄
└─ Resource group * ⓘ	DesignDatabases ⌄
	Create new
└─ Managed resource group ⓘ	Enter managed resource group name

Workspace details

Name your workspace, select a location, and choose a primary Data Lake Storage Gen2 file system to serve as the default location for logs and job output.

Workspace name *	synapsews-ptb ✓
Region *	West Europe ⌄
Select Data Lake Storage Gen2 * ⓘ	⦿ From subscription ◯ Manually via URL
└─ Account name * ⓘ	(New) synapsedatalakeptb ⌄
	Create new
└─ File system name *	(New) northwinddwh ⌄
	Create new
	☑ Assign myself the Storage Blob Data Contributor role on the Data Lake

[Review + create]   < Previous   [Next: Security >]

Figure 8.4 – Creating a Synapse workspace

If you want to create a new Synapse workspace, there are a couple of settings you need to provide. You should recognize the subscription and resource group settings from when you provisioned a SQL database and when you provisioned Cosmos DB.

6. Select your subscription and select the **DesignDatabases** resource group that you created in earlier chapters.

The **Managed resource group** setting is new compared to provisioning a SQL database or Cosmos DB. When you create a Synapse workspace, Microsoft creates some extra resources. These resources will be placed in a separate resource group. This is done automatically whether you want it or not. You do, however, have the option to specify a name for this resource group.

7. Leave the **Managed resource group** setting empty.

8. Name your workspace synapsews-ptb, replacing the last three letters with your initials to create a unique workspace name.

9. Choose the region where you also created your SQL database and Cosmos DB (or the closest to your location if you didn't work along with the book in earlier chapters).

Every Synapse Analytics workspace is attached to an Azure storage account with a container configured as a data lake. Synapse will store catalog data and metadata in this storage account. You will learn more about Azure storage accounts and data lakes in *Chapter 10, Designing and Implementing a Data Lake Using Azure Storage*. For now, you can create a new account with a new container, here called a filesystem. As you can see in *Figure 8.4*, you can attach an already-existing storage account and filesystem to Synapse if that is what you need.

10. Click on **Create new** beside the **Account name** setting and enter synapsedatalakeptb as the name of the storage account to create, and then click on **OK**.

11. Click on **Create new** beside the **File system** setting, enter northwinddwh as the name of the filesystem to create, and click on **OK**.

12. Check the checkbox before the **Assign myself the Storage Blob Data Contributor role on the Data Lake Storage Gen2 account 'synapsedatalakeptb'.** text.

Security is important whenever you store data in Azure. Microsoft makes sure Synapse can work with the created storage account. But that does not mean you can add or read files in this storage account. Ticking the last checkbox makes sure you can work with files on the storage account with the credentials you are currently logged in with.

Although you now are ready to go, let's quickly look at other settings that you can provide.

13. Click on **Next: Security**.

    You can access your SQL pools using Synapse Analytics Studio. You can start Synapse Analytics Studio from the **Synapse Analytics** blade in the Azure portal. But when you want to connect to SQL pools from Power BI, for instance, you need to have valid credentials to establish a connection. On the **Security** tab, a SQL Server administrator user is created. You can change the default name and, more importantly, provide a password instead of using an automatically created one that you do not know.

14. Enter your own name as the admin username and choose a password that you can remember.

    All data on Azure disks and in Azure storage accounts is encrypted by Azure automatically. But that is only the first layer of protection. You can add your own encryption of everything in a Synapse workspace, including, for instance, data integration artifacts, by providing your own key that you create, store, and manage in Azure Key Vault. This will also enable **Transparent Data Encryption** (TDE) to all your SQL pools. This last feature can be set later for individual SQL pools, but the entire double encryption feature can only be set at the creation time of your workspace.

    For now, encryption and key management are out of scope for this book. But in real-life projects, you cannot do without them!

    The Synapse workspace we are creating gets its own managed identity in Azure Active Directory. This identity is used, for instance, to read from and write to files in the attached storage account. This identity is created automatically and gets the required permissions on the storage account automatically as well. When you start implementing pipelines in Synapse's data integration feature, these pipelines use that same identity. Leaving the **Allow pipelines (running as workspace's system assigned identity) to access SQL pools** option checked makes sure that the integration pipelines do have access to your databases.

15. Leave the **Enable double encryption using a customer-managed key** checkbox unchecked.

16. Leave the **Allow pipelines (running as workspace's system assigned identity) to access SQL pools checkbox** checked.

17. Click on **Next: Networking**.

    Networking is, unfortunately, also outside the scope of this book. It is, however, an important part of securing your data assets and should therefore be really important in real-life deployments. If you keep the **Allow connections from all IP addresses** setting checked, you can later in this chapter connect to your SQL pools using locally installed tools such as SSMS. You can later change the "all" to specific IP addresses using firewall rules.

    By enabling managed virtual networks, Azure will create a virtual network for you. Appropriate **Network Security Groups (NSGs)** are created as well. This helps you protect against unauthorized access to the different components in your workspace while allowing communication between them. You can read more about managed virtual networks at `https://docs.microsoft.com/en-us/azure/synapse-analytics/security/synapse-workspace-managed-vnet`.

18. Accept all the default settings on the **Networking** tab and click on **Next: Tags**.

19. Click on **Next: Review and create**.

20. Click on **Create**.

    All the options on the last two tabs have been discussed already in earlier chapters so we will not repeat them here.

    After you click the **Create** button, the deployment page is shown. It may take a while for your workspace to be deployed.

21. Click on **Go to resource group** once the deployment has finished.

22. In the list of resources, click on the just-created Synapse workspace.

The Synapse blade opens. It should look (something) like *Figure 8.5*:

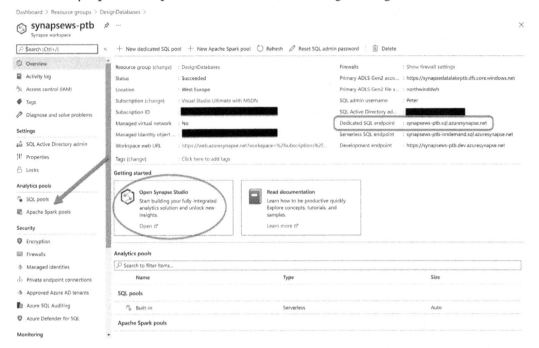

Figure 8.5 – Synapse workspace blade

A lot of options on the Synapse workspace blade are familiar from other Azure resources you may have worked with. You can find settings to configure the security using Azure Active Directory **Role-Based Access Control** (**RBAC**), change the firewall settings, and so on. *Figure 8.5* highlights a couple of items of special interest.

In the middle of the screen, you can find a button to open Synapse Analytics Studio. You will do this in the next section. On the left-hand side of the screen, in the menu, you see the Analytics pools where you can find both the SQL and Spark pools. There is always a serverless SQL pool available that cannot be deleted. The other pools are for you to create. In the upper-right corner, you will find the SQL Server name that you need when you want to connect to dedicated SQL pools from tools such as SSMS and Power BI.

Now that we have a Synapse Analytics workspace, it is time to create a dedicated SQL pool in the workspace to implement a data mart.

Creating a dedicated SQL pool

A dedicated SQL pool is the Synapse name for a database. As we explained in the *Introducing SQL pools* section, you use dedicated SQL pools to implement data marts and data warehouses. We want to implement the Sales data mart of the Northwind database. You designed this data mart in the exercise of *Chapter 7*, *Dimensional Modeling*. You will still be able to follow along if you didn't do that exercise.

The first step is to create a dedicated SQL pool. You can do that directly from the **Overview** blade of a Synapse Analytics workspace. In this book, however, we will use Synapse Studio to implement our Northwind data mart:

1. If necessary, open the Azure portal.

2. Navigate to the Synapse Analytics workspace blade.

3. Click on **Open Synapse Studio**.

    Synapse Studio will open in a new browser tab. You will get an error if your browser blocks popups and third-party cookies. You may have to log in. You can use your Azure account to log in to Synapse Studio. Once you are logged in, you will see Synapse Studio as shown in *Figure 8.6*:

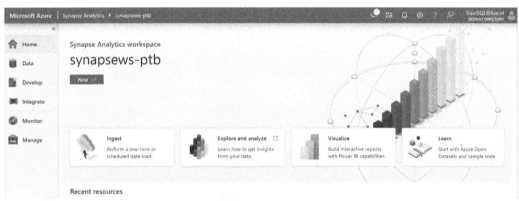

Figure 8.6 – Synapse Studio

On the left-hand side, you see the Synapse Studio menu. When collapsed, it only shows you the icons. When expanded, you can see the explanation text along with the icons. You can create a new SQL pool from the **Manage** menu option.

4. Click on the **Manage** menu option on the left-hand side of Synapse Studio.

You will now see the **Manage** blade. The first thing to do in real life is to set up source control. In the top-left corner of the **Manage** blade, you see the words **Synapse live** with a little down arrow beside it. If you click on the arrow, you will see an option to set up a code repository. You can link your Synapse Studio to Azure DevOps or a GitHub repository. For real projects, this is what you want (need) to do. All the development work that you are going to do will be safely stored in a central place, GitHub. From there, you can set up who works on this project and you can set up versioning. Once you have set up your repository, you can publish your work from Synapse Studio to the repository. Setting up GitHub, however, is beyond the scope of this book.

5. Click on **SQL Pools** in the menu of the **Manage** blade.

6. Click on **+ New**.

7. Enter NW_Sales_DM.

8. Set the performance level to DW100c (the lowest option available).

Notice that the estimated cost is shown per hour. At the time of writing, the DW100c performance tier is estimated to cost \$1.20 per hour. This is well over 800 US dollars per month. The estimated cost of a SQL database is shown in a per-month grain. The cheapest is less than 5 US dollars per month. A Synapse SQL pool is the scale-out version of a SQL database. It can handle much larger data volumes but at a higher price.

9. Click on **Next: Additional settings**.

Notice that you can create a new SQL pool from an existing backup. This is identical to Azure SQL Database. Microsoft will create restore points for any SQL pool automatically. They use snapshot technology to do so. A snapshot is taken every 8 hours and then kept for 7 days.

There is also an option to use a restore point. You might implement a strategy where you pause a SQL pool during periods of inactivity. You could, for instance, use a Synapse SQL pool to create a star schema data mart using your data lake as a source. You leverage the compute power and scale of Synapse to transform the raw data in the data lake into a star schema. You could then process an Azure Analysis Services database using the data in the SQL pool. After the Analysis Services database has been processed, you pause the SQL pool to save costs. You direct Power BI users to use the Analysis Services database. You do need to look carefully at the data usage patterns to see whether using only Analysis Services will work for you.

Using the preceding strategy, you might want to create a restore point of your SQL pool just after you have loaded new data into the SQL pool. Automatic restore points cannot be created when your SQL pool is paused. The option to create a new SQL pool from a restore point enables you to restore your SQL pool to a point in time you choose yourself.

Also notice the **Collation** setting. This is also identical to what we already know from Azure SQL Database. The collation determines whether your SQL pool is going to be case-sensitive or not and whether it is going to be accent-sensitive or not. This applies to metadata such as table names. It functions as the default for alphanumerical data types in the SQL pool. The collation also defines the code page used for non-Unicode character columns. You might want to look back at data types and collation in *Chapter 4, Provisioning and Implementing an Azure SQL DB*.

10. Click on **Next: Tags**.

11. Click on **Next: Review + create**.

12. Click on **Create**.

The last two tabs are familiar by now. It might take a couple of minutes to create your SQL pool. After the SQL pool is created, you should now see something like *Figure 8.7*:

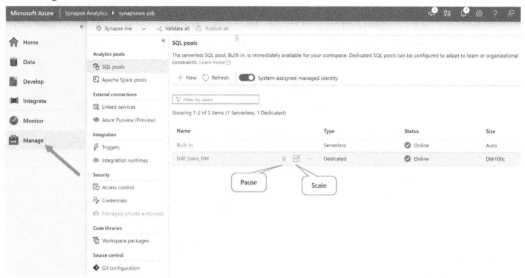

Figure 8.7 – Managing SQL pools

Notice that you have the option to pause your SQL pool, as shown in *Figure 8.7*. You only see the icons when you hover your mouse over the SQL pool. Make sure that if you're following along with the book, you pause your SQL pool every time you stop for the day or take a break. It is costly to leave a dedicated SQL pool running when you are not using it.

13. Optional: When you plan to take a break here, click on the **Pause** icon to pause the NW_Sales_DM SQL pool and click on **Pause** in the dialog that appears. Make sure that you restart your SQL Pool the next time you continue working through this chapter.

You just created a SQL pool, a scale-out SQL database to implement relational data warehouses and data marts. The next step is to create tables in your database.

Implementing tables in Synapse SQL pools

In *Chapter 4, Provisioning and Implementing an Azure SQL DB*, we started with a section on SQL Server data types, followed by a section on quantifying the data model. We should do the same here. In *Chapter 7, Dimensional Modeling*, you learned how to create a logical model of a star schema database. You will have to translate the logical model into a physical model before implementing the database. Luckily for us, Synapse Analytics SQL pools are based on SQL Server. That means we have the same data type system. You might want to re-read the section SQL Server data types of *Chapter 4, Provisioning and Implementing an Azure SQL DB*.

Quantifying the model and understanding how your SQL pool will be used is also important before implementing the tables. With a Synapse SQL pool, there is an additional design step that is *very* important. As we already said, a SQL pool is implemented by using 60 databases, called distributions. You need to decide for each table how to distribute the data in the table over the 60 distributions. This is called table geometry. Let's look at table geometries using examples:

1. If necessary, open the Azure portal. From the portal, open your Synapse workspace and open Synapse Studio from there.

2. If you paused your SQL pool earlier, resume it now. (Open the **Manage** blade in Synapse Studio and find your SQL pool. The **Pause** button, as described in the previous section, is now a **Resume** button.)

3. After the SQL pool has come online, click on the **Develop** menu option.

4. In the **Develop** blade, click on the + icon (add new resource) and click **SQL script** in the pop-up menu.

5. In the SQL script that opens, select the **NW_Sales_DM** database next to the
    Connect to option. See *Figure 8.8*:

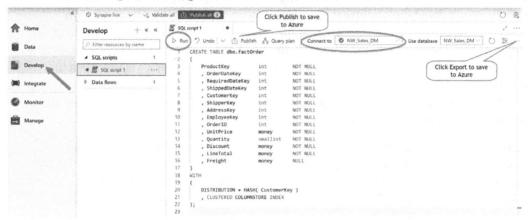

Figure 8.8 – Creating factOrder

6. Copy the following script, which creates the `factOrder` table into the script that
    opens. You can find the code in the downloads as `Create_factOrder.sql`:

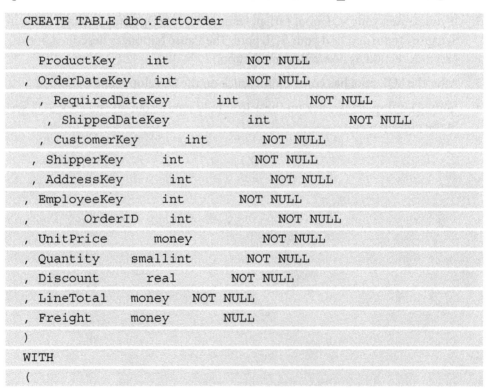

```sql
CREATE TABLE dbo.factOrder
(
    ProductKey      int             NOT NULL
,   OrderDateKey    int             NOT NULL
,   RequiredDateKey         int             NOT NULL
,   ShippedDateKey          int             NOT NULL
,   CustomerKey     int             NOT NULL
,   ShipperKey      int             NOT NULL
,   AddressKey      int             NOT NULL
,   EmployeeKey     int         NOT NULL
,       OrderID     int             NOT NULL
,   UnitPrice       money           NOT NULL
,   Quantity        smallint        NOT NULL
,   Discount        real        NOT NULL
,   LineTotal       money   NOT NULL
,   Freight         money       NULL
)
WITH
(
```

```
                        DISTRIBUTION = HASH( CustomerKey )
                , CLUSTERED COLUMNSTORE INDEX
    );
```

7. Run the script by clicking the **Run** button to create the dbo.factOrder table.

There are a couple of things to note about the previous CREATE TABLE script. First of all, the first part is really just a CREATE TABLE statement, as you learned in *Chapter 4, Provisioning and Implementing an Azure SQL DB*. You provide a two-part name for the table, meaning a schema name followed by a dot and the table name. The combination must be unique in the database. In between braces, you then enter a comma-separated list of columns, providing a column name, the data type of the column, and a specification defining whether or not the column accepts NULL as a value.

As you may guess from the column names, lines 3 to 10 in the screenshot of *Figure 8.8* define foreign keys to dimension tables. You will implement these tables later in this module. Synapse Analytics SQL pools do *not* have explicitly defined foreign keys. You cannot add a foreign key constraint to a table. Nevertheless, the first eight columns are foreign keys on a logical level. You will use them when writing joins in SQL queries you run against the SQL pool.

The OrderID column, defined in line 11, is a degenerate dimension. It can be used to analyze data to the level of individual orders. It does not reference a separate order dimension.

Lines 12 to 15 define the facts. Notice that, compared to the Northwind database created in *Chapter 4, Provisioning and Implementing an Azure SQL DB*, the Discount column has the money data type. In the operational system, Discount is stored as a percentage, while in the dimensional model we choose to store Discount as an amount. Amounts are additive and percentages need to be calculated at the report level anyway.

The most important part to note about the factOrder table, however, is the last bit. After the columns are defined, the CREATE TABLE statement continues using the WITH keyword. Within the WITH clause, the DISTRIBUTION method used to spread the data over the cluster is defined. As you can see in the script, the factOrder table uses a HASH distribution. Let's look more closely into that.

Using hash distribution

As we already mentioned several times, you need to decide how to distribute the data over the 60 distributions that together form your SQL pool. Using a hash distribution is very similar to what you learned about partitioning when implementing a Cosmos DB container in *Chapter 6, Provisioning and Implementing an Azure Cosmos DB*. In the example of the factOrder table, you see the CustomerKey column mentioned in between braces beside the DISTRIBUTION = part. This is similar to choosing the CustomerKey column to be the partition key in a Cosmos DB container.

Each time a row is entered into the factOrder table, Synapse will first look at the value of the CustomerKey column. A hash function will "translate" that value into a value between 1 and 60. This number is then used to store the new row in the corresponding distribution. A hash function guarantees that the same input always leads to the same number. All rows with the same value for CustomerKey will end up in the same distribution.

As with Cosmos, you need to choose your hash key very carefully. Suppose you only have 10 customers. That means you have only 10 different values for CustomerKey. Each customer might end up being stored in a separate distribution, but you still end up with 50 distributions that you do not use.

There are always two main factors to take into account with a database: the storage and the compute. In order to take full advantage of the 60 databases that you have inside a Synapse SQL pool, you want your data evenly distributed over all 60 databases. Each distribution should hold approximately the same number of rows. You want each database to be used equally to store the data.

Storing the data is only half the story. If you distribute your rows evenly over the cluster (the distributions) but all queries only use rows stored in a single distribution, you end up with one database doing all the work and 59 databases doing nothing. You want the compute to be evenly distributed over all distributions as well.

The best practice coming from this looks a lot like the best practice for choosing a Cosmos DB partition key.

> **Note**
>
> Always choose a hash key with a high count of distinct values that distributes the storage and compute evenly over all 60 distributions.

Aside from knowing your data and specifically knowing the number of distinct values stored in the various columns in your table, you need to know your intended workload. You most likely need to prioritize queries. When you often use a column in a SQL WHERE clause, using that column as a partition key may help improve the performance of those queries. The SQL engine will use the same hash function to determine beforehand which distribution rows are stored. It will only query that distribution, leaving all the other 59 distributions untouched.

Even more importantly than WHERE clauses are the joins that you will execute. Let's explain JOIN performance by adding another table to the SQL pool:

1. In the **Develop** blade, click on the + icon (add new resource) and click **SQL script** in the pop-up menu to add a second SQL script.

2. Make sure the new SQL script is connected to the NW_Sales_DM database.

3. Add the following code (you can find the code in the downloads in the Create_ dimCustomer.sql file):

```sql
CREATE TABLE dbo.dimCustomer
(
  CustomerKey              int             NOT NULL
, BronKey                  nchar(5)        NOT NULL
, Customer_CompanyName     nvarchar(40)    NOT NULL
, Customer_ContactName     nvarchar(30)    NULL
, Customer_ContactTitle    nvarchar(30)    NULL
, Customer_Address         nvarchar(60)    NULL
, Customer_City            nvarchar(15)    NULL
, Customer_Region          nvarchar(15)    NULL
, Customer_PostalCode      nvarchar(10)    NULL
, Customer_Country         nvarchar(15)    NULL
, Customer_Phone           nvarchar(24)    NULL
, Customer_Fax             nvarchar(24)    NULL
, StartDate                date            NOT NULL
, EndDate                  date            NOT NULL
)
```

```
WITH
(
              DISTRIBUTION = HASH( CustomerKey )
          , CLUSTERED COLUMNSTORE INDEX
);
```

4. Run the script by clicking the **Run** button to create the dbo.dimCustomer table.

One thing to note about this CREATE TABLE script is the CustomerKey column. It is used as the HASH key like in the dbo.factOrder table. It is also a surrogate key. Unfortunately, you cannot use the **Identity** property of SQL Server on columns that you use as hash keys. So, you either use a different hash key or you need an alternative to generate surrogate key values. Azure Data Factory has a surrogate key transformation in its Data Flow functionality that can help.

Let's get back to why we create a second table: the join performance in Synapse SQL pools. To understand the join performance, we need to understand the architecture of Synapse a little bit:

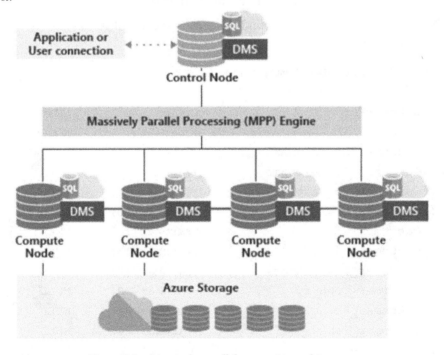

Figure 8.9 – Massively parallel processing architecture

Figure 8.9 shows how you connect to the control node of the system. The control node uses the **Massively Parallel Processing (MPP)** engine to distribute the query over all distributions. In *Figure 8.9*, the distributions are stored on four compute nodes, meaning that each compute node holds 15 distributions. This resembles a SQL pool provisioned with 2,000 DWUs. All distributions do part of the work and send their result back to the control node. The control node makes one result set of the individual intermediate results of the distributions and sends the result to the client application.

The interesting part of *Figure 8.9* is the purple boxes with DMS inside them. **DMS** stands for **data movement service**. To explain the purpose of the DMS, suppose you write a query that uses a join between the `factOrder` and `dimCustomer` tables based on the `CustomerKey` column. When both tables use the `CustomerKey` column as the hash key, it is guaranteed that rows with the same value for `CustomerKey` are stored in the same distribution. This is independent of the table; they use the same hash function. This means that each distribution can do the join on the data it holds completely independent from all other distributions. All parts of the results together form the complete result. You are using the system to its full extent.

But now suppose the `factOrder` table uses the `ProductKey` column as its hash key. A row from the `factOrder` table of a specific customer can now be on any of the 60 distributions. That makes it very likely it is not stored in the same distribution as the row of that customer in the `dimCustomer` table. The DMS kicks in and re-distributes rows from one of the tables in such a way that rows with the same value for `CustomerKey` will end up in the same distribution. Only now can the distributions start doing their part of the join. On large tables, it can take a lot of time for the DMS to do its job. Optimizing your entire workload minimizes how much the DMS needs to do.

To recap: when you join two tables on a column and both tables use that column as their hash key, each distribution can do a part of the join without the DMSes having to do anything at all. This uses the distributed architecture in the most efficient way, giving you great performance. But when you join two tables that use different columns as the hash key, the DMS makes sure you get the proper results but at the cost of lower performance.

An important note here is that you can only have one column as the hash key of a table. This means that you can optimize the table structure for a specific join. Joining the table to another table is then not optimized. However, there are other table geometries to choose from. Let's look at what a replicated table geometry means.

Using replicated distribution

Let's add a date dimension to the two tables we have so far:

1. In the **Develop** blade, click on the + icon (add new resource) and click **SQL script** in the pop-up menu to add a second SQL script.

2. Make sure the new SQL script is connected to the `NW_Sales_DM` database.

3. Add the following code (you can find the code in the downloads in the `Create_dimDate.sql` file):

```
CREATE TABLE dbo.dimDate
(
DateKey                         int             NOT NULL
, Date                          date            NOT NULL
, DayNumber                     int             NOT NULL
, Day                           nvarchar(15)    NOT NULL
, Week                          int             NOT NULL
, ISO_Week                      int             NOT NULL
, DayOfWeek                     int             NOT NULL
, CalendarMonthNumber           int             NOT NULL
, MonthKey                      int             NOT NULL
, FiscalMonthNumber             int             NOT NULL
, Month                         nvarchar(15)    NOT NULL
, MonthOfYear                   nvarchar(25)    NOT NULL
, CalendarQuarter               nvarchar(25)    NOT NULL
, FiscalQuarter                 nvarchar(25)    NOT NULL
, CalendarYear                  nvarchar(7)     NOT NULL
, FiscalYear                    nvarchar(7)     NOT NULL
, FirstDayOfMonth               date            NOT NULL
, LastDayOfYear                 date            NOT NULL
, DayOfYear                     int             NOT NULL
)
WITH
(
            DISTRIBUTION = REPLICATE
          , CLUSTERED INDEX (DateKey ASC)
);
```

4. Run the script by clicking the **Run** button to create the `dbo.dimDate` table.

As you can see in the code that creates the `dimDate` table, the distribution of this table is `REPLICATE`. Now, `REPLICATE` means that you will *not* distribute the rows of the table over the 60 distributions, but that you copy the entire table to each and every distribution. An Azure Synapse SQL pool is created for vast data volumes. But suppose you add rows for 30 years in the `dimDate` table. That is less than 11,000 rows. This is almost nothing for a Synapse SQL pool. So, although creating 60 copies of the entire table seems inefficient, you will hardly notice the extra storage needed compared to the size of the fact table.

The great thing about replicated tables is of course the join performance. Because all rows of the `dimDate` table reside in all distributions, each distribution can do part of the join between the `factOrder` fact table and the `dimDate` dimension table independently of other distributions. The DMS has no work to do to prepare for this join.

> **Note**
>
> Use `DISTRIBUTION = REPLICATE` for all small dimension tables to optimize the join performance between the fact table and the small dimension tables.

Until now, we optimized the join performance between the `factOrder` and `dimCustomer` tables by choosing the same hash key for both tables. We optimized the join performance between the `factOrder` and `dimDate` tables by replicating the `dimDate` table over all distributions. If you have multiple dimension tables that are too big to replicate, you will have to choose which join to which dimension table to optimize.

Suppose you have a `dimProduct` dimension table as well as the `dimCustomer` table. If both tables are too big in size for you to replicate them, you will have to prioritize. You can either optimize the join between the `factOrder` and `dimCustomer` tables as we did in the examples so far or you can choose the `ProductKey` column to be the hash key for the `factOrder` table instead of the `CustomerKey` column. In that case, the performance of joins between the `factOrder` and `dimProduct` tables will improve drastically at the cost of slower joins between the `factOrder` and `factCustomer` tables.

Notice that this is an economic choice. Tables using columnstore storage are unlimited in size. However, you pay for storing data. Even when you pause your Synapse SQL pool, you keep paying for the data you store inside your SQL pool. You pay more when you store more data. You need to decide what is more expensive to you: paying more for storage and having faster performance or paying less money for the Synapse pool and having less-performant queries. The same goes, of course, for the DWUs you provision. More DWUs mean more resources, which means better performance with the right database and table design at the cost of spending more money.

Next to hash and replicated distribution, Synapse SQL pools have a third option to distribute data: ROUND_ROBIN.

Using ROUND_ROBIN distribution

There might be tables that are too big to use REPLICATE for but do not have an obvious hash key to define the best distribution strategy. When all your dimensions use REPLICATE and you do not use SQL WHERE clauses on columns in the fact table (you filter on dimension attributes), there is no obvious hash key.

There are also situations where you always use all rows of a table without using joins. Staging tables are a good example of this. Data is imported from the data lake into a Synapse SQL pool table. Once imported, we can use T-SQL to prepare the data and update the production tables with this new data.

In cases like this, ROUND_ROBIN might be the best distribution strategy. All rows are randomly distributed over the 60 distributions, ensuring that all distributions get an equal part of the rows. Because rows are basically distributed randomly, there is no knowing where a specific row is stored. The system cannot use metadata when trying to find specific rows. This is a disadvantage but does not hurt performance in any way when you do not look for specific rows but always use all your rows in the queries you execute.

An advantage is that you always have an even distribution of all rows over the available distributions. Choosing the wrong hash key can lead to skew, which is when one distribution stores a lot more rows than another distribution.

Let's look at an example. Note that this is merely an example and not a well-considered design choice:

1. In the **Develop** blade, click on the + icon (add new resource) and click **SQL script** in the pop-up menu to add a new SQL script.

2. Make sure the new SQL script is connected to the NW_Sales_DM database.

3. Add the following code (you can find the code in the downloads in the `Create_dimProduct.sql` file):

```
CREATE TABLE dbo.dimProduct
(
ProductKey                      int                 NOT NULL
, SourceKey                     int                 NOT NULL
, ProductName                   nvarchar(40)        NOT NULL
, CategoryName                  nvarchar(15)        NOT NULL
, QuantityPerUnit               nvarchar(20)        NULL
, UnitPrice                     money               NULL
, ReorderLevel                  smallint            NULL
, Discontinued                  bit                 NOT NULL
, Supplier_CompanyName          nvarchar(40)        NOT NULL
  Supplier_ContactName          nvarchar(30)        NULL
, Supplier_ContactTitle         nvarchar(30)        NULL
, Supplier_Address              nvarchar(60)        NULL
, Supplier_City                 nvarchar(15)        NULL
, Supplier_Region               nvarchar(15)        NULL
, Supplier_PostalCode           nvarchar(10)        NULL
, Supplier_Country              nvarchar(15)        NULL
, Supplier_Phone                nvarchar(24)        NULL
)
WITH
(
            DISTRIBUTION = ROUND_ROBIN
          , CLUSTERED INDEX (ProductKey ASC)
);
```

4. Run the script by clicking the **Run** button to create the `dbo.dimProduct` table.

Let's shortly recap what we have learned up to here. A table in a Synapse dedicated SQL pool is created by using the T-SQL CREATE TABLE statement. It uses the same data type system as an Azure SQL database. Because Synapse Analytics is a distributed system, you need to choose a distribution strategy that defines how rows in a table are distributed over 60 distributions.

To implement a Synapse SQL pool, you take the following outlined steps:

1. Create a logical design for your database. This will often be a star schema design.

2. Transform the logical design into a physical design. Choose the proper data types. Data types have a huge impact on functionality and performance.

3. Choose the appropriate table geometry. Use REPLICATE for small tables. Use HASH distribution to optimize searches and joins between large tables. Use ROUND_ ROBIN when you always scan the entire table without the need to optimize for a specific join or WHERE clause.

4. Choose a proper index strategy.

This last bullet brings us to the next, until-now-ignored topic: the columnstore index.

Implementing columnstore indexes

If you look back at the SQL script that creates the factOrder table, you can see that CREATE TABLE ended with the WITH clause, as repeated here:

```
WITH
(
            DISTRIBUTION = HASH( CustomerKey )
    , CLUSTERED COLUMNSTORE INDEX
);
```

In the WITH clause, you can see that we explicitly defined the table to use a CLUSTERED COLUMNSTORE INDEX. You can omit this part from CREATE TABLE without changing anything because using columnstore indexes is the default in Synapse SQL pools. Let's look into what columnstore indexes are.

We explained the difference between **Online Transaction Processing (OLTP)** and **Online Analytical Processing (OLAP)** in *Chapter 1, Introduction to Databases*. OLAP describes a query workload where the majority of the queries use a lot of rows but only a few columns. A typical OLAP query might look something like the following:

```
SELECT
                CalendarYear
                , Customer_Country
                , SUM(LineTotal) AS Sales
FROM dbo.factOrder
INNER JOIN dbo.dimDate ON dimDate.DateKey = factOrder.
OrderDateKey
INNER JOIN dbo.dimCustomer ON dimCustomer.CustomerKey =
factOrder.CustomerKey
GROUP BY
                CalendarYear
                , Customer_Country;
```

This query uses all rows from the `factOrder` table to create a result set showing sales by country by year. It uses the `OrderDateKey` and `CustomerKey` columns to join the fact table to two dimensions and uses the `LineTotal` column to calculate an aggregated sales amount. All other columns of the `factOrder` table are not used. Something similar is true for both used dimension tables. In fact, typical OLAP queries, such as the one shown, use 10 to 15 percent of the columns in the table when using a table.

An Azure SQL database will, by default, store entire rows. So, all values of all columns are stored together. When you select a single value from a single column, the entire row needs to be retrieved from the database. For an OLAP query, that would mean 85 to 90 percent overhead in the data that is retrieved because that is how many column values the query will *not* use.

A columnstore index, as the name suggests, stores values not row by row but column by column. It will take all the values for the first column and store them together. It will do the same for all other columns. When a query uses just a single column, the database can retrieve just the values of that column instead of getting values for all columns. This is far more efficient than fetching entire rows. However, when you execute a query to retrieve all columns of a single row, the columnstore index all of a sudden becomes really inefficient.

So, depending on the actual workload, a columnstore or a rowstore is more efficient. But Synapse Analytics, as the name suggests, is meant to optimize analytical queries on tables with a very large number of rows. That is why the columnstore index is the default.

You can change the storage structure of a table to a rowstore if you want to. Look at the following code, which is a part of the CREATE TABLE statement we used earlier to create the dbo.dimProduct table:

```
WITH
(
            DISTRIBUTION = ROUND_ROBIN
          , CLUSTERED INDEX (ProductKey ASC)
);
```

In this code fragment, a "normal" index (or rowstore index) is created instead of a columnstore index. When creating a rowstore, you need to specify an index key. In the preceding example, the ProductKey column is used as the index key. If you create a pop-up window in Power BI using DirectQuery that shows all the details of a product when you hover your mouse over a product name, this index might be beneficial.

We could write an entire book on the subject of indexing alone. An in-depth discussion is beyond the scope of this book. There is, however, one more important note to make about creating columnstore indexes. The query that creates the index, or queries that bulk-load data into tables that use the clustered columnstore index storage structure, needs a lot of memory. You can make sure enough memory is available by assigning to queries resources using workload groups. That is what the next section is about.

Understanding workload management

A Synapse Analytics SQL pool is a scale-out SQL implementation. Even though it is a scale-out system, resources might be scarce and they are expensive. You need to be able to regulate how your resources are utilized by your queries. Assigning resources to queries is called workload management.

There is a trade-off to consider. A Synapse SQL pool can today execute 128 concurrent queries. If you do have 128 concurrent queries, all available resources are divided over those queries, which means that each query only has a limited number of resources it can use. Resources are a combination of I/O, memory, and CPU. When bulk-loading a columnstore table, you need your query to be able to utilize a lot of memory. With too many concurrent queries, your bulk-load query might not have enough memory. You can assign more resources to a query but that means you can run fewer queries concurrently.

To understand the memory requirement of bulk-load queries, we need to understand columnstore indexes a bit better. When you add a new row to a columnstore table, it is added as a row. In an optimal situation, Synapse waits for 1,048,576 new rows to arrive. Once it has 1,048,576 new rows, it will transform the data into the columnstore storage format. Microsoft uses the term "compression" to describe this process. The compression process starts when you either have 1,048,576 new rows or your query runs out of memory, whatever comes first. Because storage and compression take place on individual distributions, you have to multiply these numbers by 60. You want to load your columnstore tables in batches of 60 times 1,048,576 rows to get optimal columnstore storage. When you start compressing data with fewer rows, either because you have smaller batches defined or your query is out of memory, you get index fragmentation. Fragmentation will impair query performance. The conclusion is that a query loading data should get assigned enough memory to hold 60 times 1,048,576 rows in memory.

Workload management is the task to make sure your query gets enough memory. Workload management is done in two steps. First, you create one or more workload groups that you assign resources to. Then you create so-called classifiers that assign queries to a workload group to run in. The query can then use the resources assigned to the workload group.

Creating a workload group

The first step is to create a workload group:

1. In the **Develop** blade, click on the + icon (add new resource) and click **SQL script** in the pop-up menu to add a new SQL script.

2. Make sure the new SQL script is connected to the NW_Sales_DM database.

3. Add the following code (you can find the code in the downloads in the Workload_Management.sql file):

```
CREATE WORKLOAD GROUP Batch_Loads
WITH
(
```

```
    MIN_PERCENTAGE_RESOURCE = 25
    , CAP_PERCENTAGE_RESOURCE = 25
    , REQUEST_MIN_RESOURCE_GRANT_PERCENT = 25
    , REQUEST_MAX_RESOURCE_GRANT_PERCENT = 25
    , IMPORTANCE = HIGH
    , QUERY_EXECUTION_TIMEOUT_SEC = 0
);
```

4. Run the script by clicking the **Run** button to create the `Batch_Loads` workload group.

Then preceding script creates a workload group called `Batch_Loads`. Let's have a look at what all the settings mean.

`MIN_PERCENTAGE_RESOURCE` defines how many resources the system will always assign to this workload group. These resources are guaranteed to be available for queries running inside this workload group. The values of all workload groups together cannot exceed 100 percent. You cannot just choose values, however. Each query needs a minimum number of resources to execute. Depending on the service level you provisioned, that means there is a minimum value you can set for `MIN_PERCENTAGE_RESOURCE`. If you specify a lower value, the system will make the value 0. If you scale your system to a higher service level, you may find that your configured value is now above the minimum value allowed and will take effect. This means that scaling has an effect on how queries get assigned resources. This might make it difficult to foresee exactly what happens to which queries when you decide to scale up your SQL pool.

`CAP_PERCENTAGE_RESOURCE` defines how many resources all queries running concurrently in this workload group may use together. When new requests arrive and there are not enough resources available, the new request will be queued until enough resources become available. If you create four workload groups with a `MIN_PERCENTAGE_RESOURCE` setting of 25 each, the `CAP_PERCENTAGE_RESOURCE` setting will automatically be reduced to 25 as well because the `MIN_PERCENTAGE_RESOURCE` setting guarantees the resources are available for the other workload groups.

REQUEST_MIN_RESOURCE_GRANT_PERCENT has a value between 0.75 and 100 and is always a multiple of 0.25 and a fraction of MIN_PERCENTAGE_RESOURCE. It defines the minimum amount of resources available for an individual request. In our example, we have REQUEST_MIN_RESOURCE_GRANT_PERCENT at a value of 25, which is exactly the same as the setting we used for MIN_PERCENTAGE_RESOURCE. One query executing in the context of this workload group can use 25 percent of the overall resources available, which is 100 percent of the resources allocated to this particular workload group. That means that a maximum of one query can execute in this workload group at any point in time.

We do not have a lot of options in our example because we provisioned a DW100c service tier. Each query needs a minimum of 25 percent of the resources. No matter how you create your workload groups, you will never have more than four concurrent queries. Service levels of DW6000c and higher can have 128 concurrent queries if you specify the REQUEST_MIN_RESOURCE_GRANT_PERCENT setting to be 0.75 percent.

REQUEST_MAX_RESOURCE_GRANT_PERCENT allocates extra resources to a query to a maximum of the specified value when those resources are available because no other concurrent queries use them.

The IMPORTANCE setting defines the default importance for a query executed in the context of this workload group. When the sum of all MIN_PERCENTAGE_RESOURCE settings is less than 100, there are non-reserved resources. These resources are assigned to queries based on importance. The importance levels you can choose from are LOW, BELOW_NORMAL, NORMAL (default), ABOVE_NORMAL, and HIGH.

The last setting speaks for itself: QUERY_EXECUTION_TIMEOUT_SEC. A query is canceled when its execution time exceeds the specified number of seconds. A value of 0 means that queries never get canceled but can run until they finish no matter how long that takes.

Only the MIN_PERCENTAGE_RESOURCE, CAP_PERCENTAGE_RESOURCE, and REQUEST_MIN_RESOURCE_GRANT_PERCENT settings are mandatory.

Now that we have a workload group, it is time to create a classifier.

Creating a workload classifier

Let's get straight into creating a classifier:

1. In the **Develop** blade, click on the + icon (add new resource) and click **SQL script** in the pop-up menu to add a new SQL script.

2. Make sure the new SQL script is connected to the NW_Sales_DM database.

3. Add the following code (you can find the code in the downloads in the Workload_ Management.sql file):

```
CREATE WORKLOAD CLASSIFIER Classify_Batch_Loads
WITH
(
        WORKLOAD_GROUP        = 'Batch_Loads'
        , MEMBERNAME                  = 'dbo'
-- Optional extra settings
        --, IMPORTANCE       = NORMAL
        --, WLM_LABEL        = 'label'
        --, WLM_CONTEXT      = 'name'
        --, START_TIME       = 'start_time'
        --, END_TIME         = 'end_time'
);
```

4. Run the script by clicking the **Run** button to create the Classify_Batch_Loads workload classifier.

A workload classifier adds a query request to a workload group based on several criteria. In the preceding example, queries executed by the dbo user are classified to execute in the Batch_Loads workload group. For the CREATE statement to execute successfully, you have to enter an existing database user, database role, Azure Active Directory login, or Azure Active Directory group for the MEMBERNAME setting. In a real-life situation, you would create a database role or an Azure Active Directory user called something such as ETL_User that you can use to execute the ETL workload and use to classify against. In this example, we simply used dbo because it already exists.

Apart from the MEMBERNAME setting, you can use any of the other criteria to classify queries. You can, for example, add an OPTION clause to queries to provide a query with a label. You can see an example in the code shown here:

```
SELECT COUNT(*)
FROM DimCustomer
OPTION (LABEL = 'dimension_loads');
```

You can create a workload classifier using the same label, dimension_loads. Another option is to set a context for an entire session instead of providing a label per query. You do this using the following stored procedure:

```
 EXEC sys.sp_set_session_context @key = 'wlm_context', @value =
'dim_load';
```

You can classify this session using a classifier that uses dim_load for the WML_CONTEXT setting. The last option is to use time periods. For instance, ETL processes might run during the night and need a lot of resources. You can schedule your ETL using Azure Data Factory in such a way that you do not need a lot of concurrency during the night. During the day, you might want to facilitate business users. Their queries need fewer resources but you need more concurrency in the system.

You might have two or more classifiers that satisfy a request. For instance, one classifier specifies the user trying to execute a query while another classifier has a period defined and the query runs at a time within that defined period. In this case, the classifier based on the user takes precedence. The order of precedence is user, role, WML_LABEL, WLM_CONTEXT, START_TIME/END_TIME.

Now that we have tables and know how to use our resources efficiently, it is time to add some data to our tables.

Using PolyBase to load data

There are multiple ways to add data to a Synapse Analytics dedicated SQL pool. The recommended way that provides the best performance is by using PolyBase. PolyBase is a feature that enables you to write T-SQL queries in Synapse to query data that is stored in databases other than your Synapse SQL pool. There are multiple databases that PolyBase can read data from, but the most obvious one is reading data from a data lake. In Azure, we implement data lakes in Azure Storage. You will learn about that in *Chapter 10, Designing and Implementing a Data Lake Using Azure Storage*. For now, we will start by uploading data to the data lake account you created in the *Provisioning a Synapse Analytics workspace* section:

1. In the Azure portal, open the menu of the portal on the left-hand side of the screen and click on **Resource groups**.

2. Click on the resource group you created for this book (**DesignDatabases** if you followed the steps as described in the book).

3. Click, in the list of resources, on the name of the storage account listed (it should be named something like **synapsedatalakeptb**, where the last three characters are your initials instead of mine).

4. Click on **Open in Explorer** at the top of the page.

5. Download and install the Azure Storage Explorer if you haven't done so already. Otherwise, click on **Open Azure Storage Explorer**. (Installing the Azure Storage Explorer is a straightforward process.)

6. Click on **Open** when the browser tells you that the site is trying to open Microsoft Azure Storage Explorer.

7. In the Storage Explorer, select the blob container called **northwinddwh** under the storage account that was created during the creation of the Synapse Analytics workspace. See *Figure 8.10*:

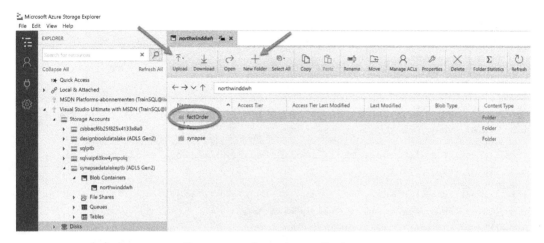

Figure 8.10 – Azure Storage Explorer

8.   Click on **New Folder**. Name the folder `factOrder` and click on **OK**.

9.   Double-click on the newly created folder.

10. Click on **Upload**, select **Upload Files…**, and upload the `factOrder_data.csv` file, which you can find in the downloads for this book.

11. Click on **Upload** to close the dialog.

Now that we have data available in our storage account, the first thing to do is make sure our SQL pool has access to this storage account. There are a couple of options to do this. The best option is creating an Azure Active Directory application that we can use to provide a SQL pool with access to a storage account.

Enabling a SQL pool to access a data lake account

The first step is to create an application registration in Azure Active Directory:

1.   In the Azure portal, open the menu of the portal on the left-hand side of the screen and click on **Azure Active Directory**.

2.   On the left-hand side of the screen, in the menu of the **Active Directory** blade, click on **App registrations**.

3.   At the top of the screen, click on **+ New registration**.

4.   Enter `DLAccess` as the name.

5. Enter `http://localhost` as the redirect URI, leaving the **Web** option selected. Your dialog should now resemble *Figure 8.11*:

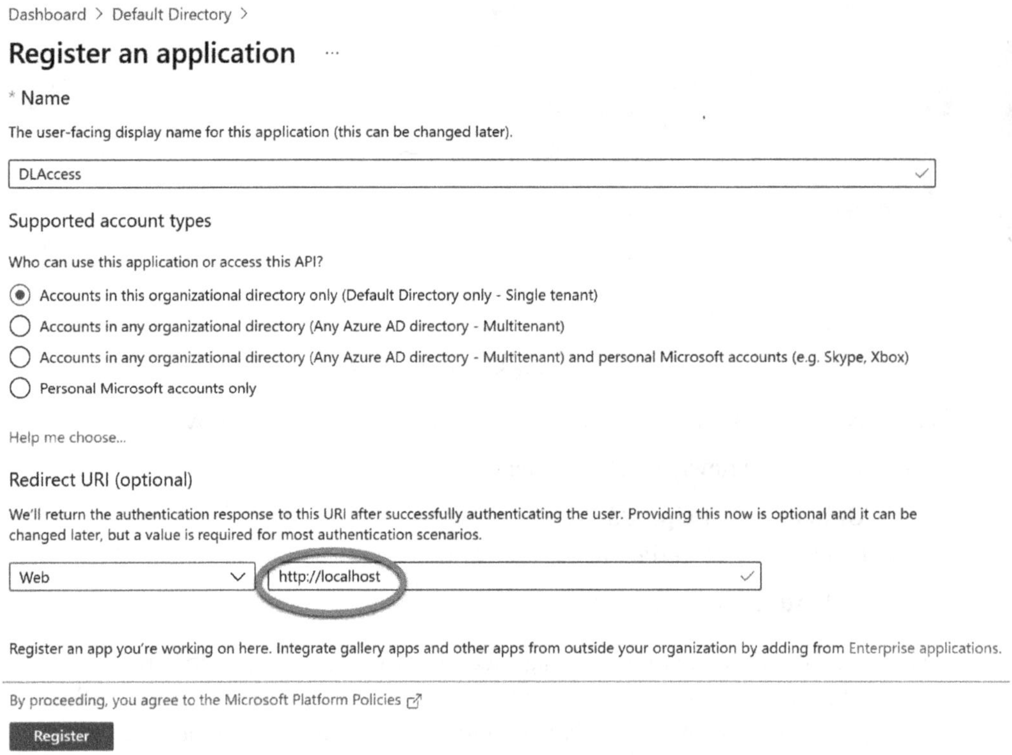

Dashboard > Default Directory >

Register an application   ⋯

* Name

The user-facing display name for this application (this can be changed later).

DLAccess	✓

Supported account types

Who can use this application or access this API?

(•) Accounts in this organizational directory only (Default Directory only - Single tenant)

( ) Accounts in any organizational directory (Any Azure AD directory - Multitenant)

( ) Accounts in any organizational directory (Any Azure AD directory - Multitenant) and personal Microsoft accounts (e.g. Skype, Xbox)

( ) Personal Microsoft accounts only

Help me choose...

Redirect URI (optional)

We'll return the authentication response to this URI after successfully authenticating the user. Providing this now is optional and it can be changed later, but a value is required for most authentication scenarios.

Web ∨	http://localhost	✓

Register an app you're working on here. Integrate gallery apps and other apps from outside your organization by adding from Enterprise applications.

By proceeding, you agree to the Microsoft Platform Policies ⌐⫟

[Register]

Figure 8.11 – Registering an application

6. On the **DLAccess** registered app screen, take a note of the application (client) ID. Copy it to a Notepad file.

7. At the top of the screen, click on **Endpoints**.

8. Copy the **OAuth 2.0 token endpoint (v1)** value. Paste it in the Notepad file of *step 6*.

9. Click, in the menu of the **DLAccess** blade, on the **Certificates & secrets** option.

10. Click on + **New Client Secret** at the top of the page.

11. Enter the **DLAccess** key as the description.

12. Click **Add**.

13. Copy the value of the key that you just created. Note that this is the only change to copy that key. When you leave this page, it will not show the key anymore, even when you return to the page. Paste the key value to the Notepad file you opened in *step 6*.

You just created an application in Azure Active Directory. You can now provide this application access to your data lake:

1. Open the menu of the portal on the left-hand side of the screen and click on **Resource groups**.

2. Click on the resource group you created for this book (**DesignDatabases** if you followed the steps as described in the book).

3. Click, in the list of resources, on the name of the storage account listed (it should be named something like **synapsedatalakeptb**, where the last three characters are your initials instead of mine).

4. Click, in the menu of the storage account on the left-hand side of the screen, on **Access Control (IAM)**.

5. Click on **+ Add** In the menu on top of the page and select **Add role assignment** in the dropdown.

   N.B.: At the time of writing, the new **Add role assignment** dialog is in preview. By the time you read this, the interface might have changed compared to how it is described here.

6. Select **Storage Blob Data Contributor** from the **Role** drop-down list.

7. Type DLAccess in the **Select** text box.

8. Click on the **DLAccess** application that appears.

9. Click on **Save**.

You just provided the application with the permissions belonging to the **Storage Blob Data Contributor** role. Maybe just read permissions would have been sufficient. Make sure you always use the least amount of necessary privileges.

Now that we have an application that has access to our data lake, it is time to configure PolyBase in our Synapse SQL pool.

Configuring and using PolyBase

The first step is to allow our Synapse SQL pool to use the registered app to authenticate to Azure Storage. You do that by creating a database scoped credential. This database object uses the application (client) ID and the secret key of the registered app. Because this is security-sensitive information, the database needs to store these values encrypted. It needs an encryption key to do that. So, the first step is to create a master key, followed by creating a database scoped credential:

1. Open Synapse Studio.

2. In the **Develop** blade, click on the + icon (add new resource) and click **SQL script** in the pop-up menu to add a new SQL script.

3. Make sure the new SQL script is connected to the NW_Sales_DM database.

4. Add the following code (you can find the code in the downloads in the Configure_Polybase.sql file). Make sure to substitute your values into the code between the < and > characters. You should use the values you copied to Notepad in *steps 6, 8,* and *13* of the previous section:

```
CREATE MASTER KEY ENCRYPTION BY PASSWORD =
'Th1s1sR3aIIyS3cr3t';


CREATE DATABASE SCOPED CREDENTIAL
AzureStorageAccountCredential
WITH
    IDENTITY = '<Application (client) ID>@<OAuth 2.0
token endpoint (v1)>,
    SECRET = '<client_secret>'
;
```

The next step is to tell the SQL pool where our data lake resides. You do that by creating an external data source.

5. Add the following code to the script of the previous step:

```
CREATE EXTERNAL DATA SOURCE AzureStorage
WITH (
    TYPE = HADOOP,
    LOCATION = 'abfss://northwinddwh@synapsedatalakeptb.
blob.core.windows.net',
    CREDENTIAL = AzureStorageAccountCredential
);
```

Make sure you get the location right. The URL uses a container named northwinddwh if you followed along with the steps as described in the book. The part after the @ sign is the name of your storage account, which should end with your initials.

There are a few things to notice in this statement. The first thing is the TYPE setting, which is set to HADOOP. Always use Hadoop if you want to read data from Azure Blob Storage, Azure Data Lake Gen 1, or Azure Data Lake Gen 2. In our current example, we use Azure Data Lake Gen 2. LOCATION uses a URL prefix of abfss. This is the prefix you use when reading from Azure Data Lake Gen 2. For a normal blob container, you would use wasbs, and for Azure Data Lake Gen 1, you would use adl as the prefix. The last thing to note is the reference to the already-created database scoped credential in the CREDENTIAL setting.

Your Synapse pool now knows where to find the data lake and how to authenticate to it. But what types of data will it find in that location? You need to tell your SQL pool what to expect by creating an external file format.

6. Add the following code to the script of the previous step:

```
CREATE EXTERNAL FILE FORMAT TextFile
WITH
(
    FORMAT_TYPE = DelimitedText
    , FORMAT_OPTIONS (FIELD_TERMINATOR = ','
                    , FIRST_ROW = 2)
);
```

There are four options available for the FORMAT_TYPE setting. You specify either Parquet, ORC, RCFile, or DelimitedText. You will learn about these file types in *Chapter 10, Designing and Implementing a Data Lake Using Azure Storage.* In our current example, we have an uncompressed CSV file that we want to read. That is why the FORMAT_TYPE setting has DelimitedText specified. If you want to read from compressed files, you need to add an extra setting to the CREATE EXTERNAL FILE FORMAT statement: DATA COMPRESSION. This is where you tell PolyBase how to decompress the data.

There are a couple of formatting options that you can specify. In our example, we tell PolyBase that we have commas as column (field) separators and that it should skip the first row because it contains column names.

Each time your data lake gets new file types, you create new external file format objects. The rest of the preceding code has to be executed once only.

7. Run the entire SQL script, creating the master key, the database scoped credential, the external data source, and the external file format.

There remains one thing to be done before we can write T-SQL code to query the data in the CSV file in the data lake. There is no such thing as SQL without metadata. You can provide PolyBase with the required metadata by creating an external table:

1. In the **Develop** blade, click on the + icon (add new resource) and click **SQL script** in the pop-up menu to add a new SQL script.

2. Make sure the new SQL script is connected to the NW_Sales_DM database.

3. Add the following code (you can find the code in the downloads in the Use_ Polybase.sql file):

```
CREATE EXTERNAL TABLE dbo.ext_factOrder
(
ProductKey                      int
, OrderDatumKey                 int
, RequiredDatumKey              int
, ShippedDatumKey               int
, CustomerKey                   int
, ShipperKey                    int
, AddressKey                    int
, EmployeeKey                   int
, OrderID                       int
```

```
  , UnitPrice                          money
  , Quantity                           smallint
  , Discount                           money
  , LineTotal                          money
  , Freight                            money
  )
WITH
  (
       LOCATION='/factOrder/factOrder_data.csv',
       DATA_SOURCE=AzureStorage,
       FILE_FORMAT=TextFile
  );
```

4. Run the script by clicking the **Run** button to create the dbo.ext_factOrder
    table.

An external table is like any table in the sense that you can write SQL statements against
it. As a query developer, you need not even know whether a table is an external table or
an ordinary table to write SELECT statements using the table. An external table, however,
is more like a view. It does not store any data by itself. The last part of the preceding
statement references the factOrder_data.csv file that we uploaded at the beginning
of this section. The data is in the file; the metadata describing the data is in our SQL pool.
When you drop external data, you do not lose any data. The underlying file(s) will still be
present in your data lake.

You can test the external table by querying it. Open a new SQL script and enter, for
instance, the following query:

```
SELECT COUNT(*) FROM dbo.ext_factOrder;
```

The last step to undertake is to import the data and populate our dbo.factOrder table
with the data from the file. We can do that using what is called the CTAS statement.

Using CTAS to import data

CTAS stands for **Create Table As Select** and is a preferred way to import data into dedicated SQL pools. A CTAS statement creates a new table and populates that table with data based on a SELECT statement:

1. Open a new SQL script and enter the following code (you can find the code in the CTAS.sql file):

    ```
    CREATE TABLE dbo.stage_factOrder
    WITH
    (
        DISTRIBUTION = ROUND_ROBIN
        , CLUSTERED COLUMNSTORE INDEX

    )
    AS
    SELECT * FROM dbo.ext_factOrder;
    ```

2. Execute the CTAS statement.

    The preceding statement creates a new table called dbo.stage_factOrder. You will get an error if a table with that name already exists. Notice that we used a ROUND_ROBIN distribution. We want to utilize the entire cluster and do not plan to use any joins on this data.

    Also notice the simple SELECT statement that is the last line of the preceding code. The beauty of the CTAS statement is that you can use all known functionality of the SELECT statement. You can use WHERE clauses, JOIN clauses, computed columns, and so on. The column names of the columns that you select together with the data types of those columns define the newly created table. In our simple example, the dbo.stage_factOrder table has an identical column structure as the external dbo.ext_factOrder table. The difference is that after running the CTAS statement, the data now actually resides in the dedicated SQL pool. Querying normal tables will give you far better performance than querying external tables.

An often-utilized strategy is to import data into temporary staging tables. Once the data is loaded into staging tables, you can transform the data as needed and then load it into the production tables. In our simple example, we have to insert the data in the dbo.factOrder fact table.

3. Add the following INSERT statement to your SQL script and execute the statement (it is part of the code in the CTAS.sql file):

```
INSERT dbo.factOrder
(
            ProductKey
        , OrderDateKey , RequiredDateKey ,
ShippedDateKey , CustomerKey , ShipperKey , AddressKey ,
EmployeeKey , OrderID , UnitPrice , Quantity , Discount ,
LineTotal , Freight
)


SELECT
            ProductKey , OrderDateKey , RequiredDateKey
    , ISNULL(ShippedDateKey, '99991231') , CustomerKey
    , ShipperKey , AddressKey , EmployeeKey , OrderID ,
UnitPrice , Quantity , Discount , LineTotal , Freight
    FROM dbo.stage_factOrder;
```

Notice this statement is a regular INSERT statement, which we have been using in SQL Server for years. Also notice the ISNULL(ShippedDateKey, '99991231') line. This is a simple example of transformations that you can do. There are some rows without ShippedDateKey. That column is, however, mandatory in our fact table. So, the code substitutes the NULL value with 99991231.

With the data in our actual fact table, there are two things left to do: first, we can drop the staging table. The staging table is no longer needed. We might recreate and drop it again the next time we load new data into the fact table. We also need to create some statistics to improve the query performance of the dbo.factOrder table.

4. Add the following code to your SQL script and execute the statement (it is part of the code in the CTAS.sql file):

```
DROP TABLE dbo.stage_factOrder;
```

```
CREATE STATISTICS [stats_CustomerKey] ON [factOrder]
([CustomerKey]);
CREATE STATISTICS [stats_ProductKey] ON [factOrder]
([ProductKey]);
CREATE STATISTICS [stats_OrderDateKey] ON [factOrder]
([OrderDateKey]);
```

Statistics are really important to SQL Server query performance and that is no different for Synapse dedicated SQL pools. The query engine uses statistics to optimize queries. Bad or missing statistics will almost always lead to bad performance because inefficient query plans will be used. Statistics tell the query engine what the distribution is of the actual values in a column. The query engine uses this information to estimate how many rows are involved in each step it takes during query processing.

Like a "normal" SQL database, Synapse will automatically create statistics unless you turn the setting AUTO_CREATE_STATISTICS to OFF. Synapse will not create statistics on external tables. You can still create your own statistics by executing CREATE STATISTICS statements as in the preceding code.

Once statistics have been created, you need to keep them up to date. A best practice is to update statistics after loading new data. You do that by running the UPDATE STATISTICS statement after each time you load new data into a table. The UPDATE STATISTICS statement is equal in syntax to the CREATE STATISTICS statement.

Instead of using CTAS, we could have used the COPY statement. It provides less flexibility but also involves fewer steps.

Using COPY to import data

Synapse Studio can generate the required SQL statement for us. We will copy data to the dbo.dimDate table to populate the date dimension. We need data to copy in the data lake first:

1. Open Azure Storage Explorer.

    Note that Synapse Studio provides access to the data lake under the **Linked services** tab. You can use this just as well instead of the Azure Storage Explorer. One advantage would be that you don't need a locally installed app. You can do everything from one environment.

2. Create a new folder called dimDate in the container and storage account created with the Synapse Analytics workspace.

3. Upload the `dimDate_data.csv` file to this new folder. You can find the file in the downloads.

   With the data available, we can start importing it into our table.

4. Click, in the Synapse Studio menu on the left-hand side of the screen, on **Data**.

5. Under **Data**, click on **Workspace**, and then click on **Databases** to expand the list with available databases.

6. Click on the **NW_Sales_DM** database to expand the menu.

7. Open the folders for **Tables** and **External Tables** to verify all your tables are available.

8. Hover with the mouse over the **dbo.dimDate** table and click on the ellipsis that appears when you do.

9. Select **New SQL script** and then **Bulk load** (see *Figure 8.12*).

10. Under the **Storage Account** option, select the storage account belonging to the Synapse Analytics workspace. In my case, that is `synapsews-ptb-WorkspaceDefault`. You will have different initials in your workspace name.

11. Click on **Browse** beside the **Input file or folder** setting.

12. Double-click on the **northwinddwh** container name.

13. Double-click on the **dimDate** folder.

14. Select the **dimDate_data.csv** file and click **OK**.

15. Click **Continue**.

16. Note that you have to provide metadata on the file as if you are creating an external file format. Accept all the defaults and click on **Continue**.

17. Select the **Automatically** option and click on **Open Script**.

18. You should see something similar to *Figure 8.12*:

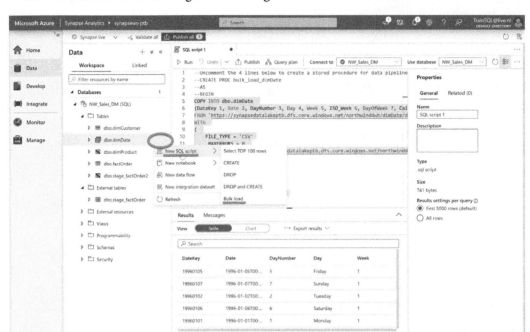

Figure 8.12 – Synapse Studio Data Explorer

You just used the COPY statement to add data to your table. It is time to see some results from all the work we did.

Connecting to and using a dedicated SQL pool

Until now, we have interacted with the SQL pool using Synapse Studio. But a dedicated SQL pool is like any SQL database. You can connect to it using familiar tools such as SSMS, Azure Data Studio, and Power BI.

Working with Azure Data Studio

Let's take a quick look by connecting to the SQL pool using Azure Data Studio:

1. Start Azure Data Studio.

2. Click on the **Connections** menu in the upper-left corner of Azure Data Studio.

3. Click on the **New connection** icon.

4. In the **Connection Details** pane on the right-hand side of your screen, select **Microsoft SQL Server**.

5. Copy the name of your server to the **Server** setting. You can find this name on the **Overview** blade of your Synapse workspace in the Azure portal. It should look something like `synapsews-ptb.sql.azuresynapse.net`, where you used your initials instead of `ptb`.

6. Choose **SQL Login** for the **Authentication type** setting and enter the username and password you created in *step 14* of the *Provisioning a Synapse Analytics workspace* section.

7. Check the **Remember password** box.

8. Enter `NW_Sales_DM` as the name of the database.

9. Click on **Connect**.

10. Right-click on the connection and select **New Query**.

11. Enter and execute the following query:

```
SELECT
      CalendarYear
    , SUM(LineTotal)
    , SUM(Discount)
FROM dbo.factOrder
INNER JOIN dbo.dimDate ON dimDate.DateKey = factOrder.
OrderDateKey
GROUP BY CalendarYear;
```

You should see what is shown in *Figure 8.13*:

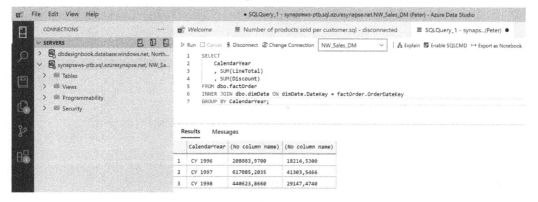

Figure 8.13 – Running queries using Azure Data Studio

Business users probably do not use Azure Data Studio but rather a tool such as Power BI.

Working with Power BI

Working with a Synapse SQL pool from Power BI is as easy as we just saw for Azure Data Studio:

1. Start Power BI Desktop.

2. Click on **Get Data** on the **Home** ribbon and select **SQL Server** from the drop-down list that appears.

3. Use the same server name as in *step 5* of the *Working with Azure Data Studio* section. It should be similar to `synapsews-ptb.sql.azuresynapse.net`.

4. Enter `NW_Sales_DM` beside the **Database** option.

5. Select the **DirectQuery** option.

6. Click on **OK**.

7. In the security dialog that appears, click, on the left-hand side, on **Database**.

8. Enter the username and password you created in *step 14* of the *Provisioning a Synapse Analytics workspace* section.

9. Click on **Connect**.

10. On the **Navigator** screen that appears, select all tables except the **ext_factOrder** table and click on **Load**.

11. From the list of fields on the right-hand side of your Power BI screen, drag the **LineTotal** field from the **factOrder** table to the middle of the screen (the report canvas).

12. Drag the **CalendarYear** field from the **dimDate** table to the bar chart that was created in the previous step.

    Unfortunately, we were not able to create foreign keys in a Synapse SQL pool. Because of that, Power BI does not know how to show `LineTotal` per year. That is, however, easily fixed in Power BI.

13. On the left-hand side of the screen, click on the icon of an ERD to open the **Model** page of Power BI.

14. Drag the **DateKey** column from the **dimDate** table and drop it on the **OrderDateKey** column in the **factOrder** table.

15. Click **OK** in the **Create relationship** dialog.

16. On the left-hand side of the screen, click on the icon of a bar chart to open the **Report** page of Power BI. You should now see what is shown in *Figure 8.14*:

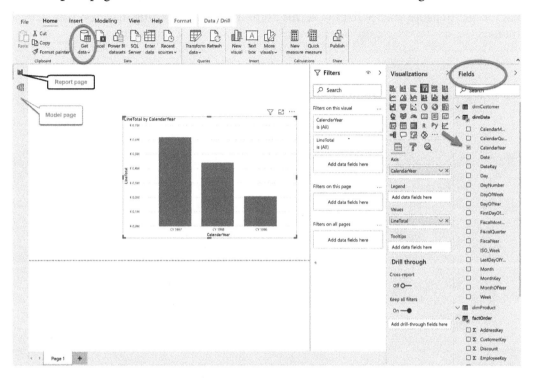

Figure 8.14 – Using Power BI

Having connected with Power BI, you can now start exploring your data.

Summary

In this chapter, you learned what Synapse Analytics is and what components it comprises. You provisioned a Synapse Analytics workspace and a dedicated SQL pool in it. You created some tables and learned about table geometries and how to choose the correct geometry for your tables. You then populated some tables with data stored in the data lake. We will continue with data loading in *Chapter 11, Implementing ETL Using Azure Data Factory*.

There is more to dedicated SQL pools than what we can explain in one chapter. We encourage you to go through the official documentation at docs.microsoft.com to learn about all the available features.

We will continue this book by learning about Data Vault in the next chapter, *Chapter 9, Data Vault Modeling*.

Do not forget to pause your SQL pool at this point. Keeping it running may cost you a lot of money.

9
Data Vault Modeling

In *Chapter 7, Dimensional Modeling*, you learned how to create a star schema database. In *Chapter 8, Provisioning and Implementing an Azure Synapse SQL Pool*, you implemented a star schema. The goal was to optimize the database for analysis and reporting.

Demands and regulations change over time, meaning that the way we use data also changes over time. These changes the way we use data for us to implement changes in the star schema designs we create. To create a stable data platform, we implement a layer in between the operational databases that we learned about in chapters 3 to 6, and the dimensionally modeled databases that we learned about in *Chapter 7, Dimensional Modeling*, and *Chapter 8, Provisioning and Implementing an Azure Synapse SQL Pool*. This layer should be optimized for the long-term, flexible storage of historical data. When data here means relational data, creating a data vault data warehouse is a good option. That is what this chapter is all about. When data also means big data, creating a data lake, as discussed in *Chapter 10, Designing and Implementing a Data Lake Using Azure Storage*, is a good choice.

This chapter is all about Data Vault modeling. Specifically, you will learn about the following topics:

- Background to Data Vault modeling
- Designing hubs, links, and satellites
- Using hash keys
- Designing a Data Vault structure

- Designing business vaults

- Implementing a Data Vault

Background to Data Vault modeling

We have already discussed nomalizing data and dimensional modeling as possible ways to design relational databases. Normalizing is a great strategy for databases that support line-of-business applications. Dimensionally modeled databases are optimized for reporting. In this chapter, we will look at a third alternative – Data Vault modeling. It optimizes the model for the flexible long-term storage of historical data.

Have a look at *Figure 9.1*, which is a copy of *Figure 7.2*:

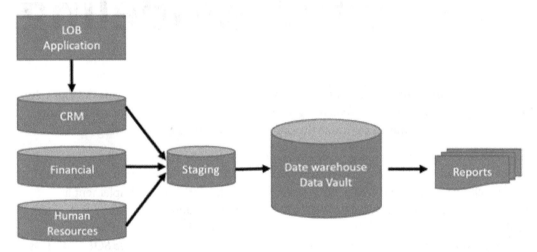

Figure 9.1 – Classic data warehouse architecture

This figure shows how Ralph Kimball envisioned the data warehouse. The data warehouse is a collection of all the star schemas describing an organization. It has a star for **Sales**, a star for **Marketing**, a star for **Human Resources**, and so forth. It is loaded from the data that originates from line-of-business applications. Whenever two processes use the same dimension, you create a conformed dimension. This is a dimension that is generic in order to be useful for all the different processes that use it.

Bill Inmon criticized this approach. He based his criticism on the business intelligence life cycle. The business intelligence life cycle states that there will be constant change in the use of your data. Have a look at *Figure 9.2*:

Figure 9.2 – Business intelligence life cycle

People analyze the data we have. By analyzing the data, they reach new insights. These insights might lead to changes to improve current business processes. Following the change, we measure the new process by collecting data. We analyze this data. Getting actionable insight is what business intelligence is all about.

Suppose, for instance, you are an IT course provider. Account managers sell courses to large companies. Each (potential) large account has a dedicated account manager assigned to it. This means that the account manager becomes a column in a flattened customer dimension. After analyzing the sales data, you find that almost all customers only buy a limited selection of all the courses you sell. When you compare a customer to a comparable customer operating in the same branch that has a different account manager assigned to it, you find that the other customer buys different courses, but also a limited selection.

Account managers have a comfort zone and tend to only sell courses that they feel comfortable with. For instance, an account manager with a background in cloud computing sells Azure and AWS courses but neglects Microsoft Office courses. With this, you miss out on a lot of potential revenue.

You decide to reorganize to increase sales. Account managers are assigned a certain portfolio. They need to spot opportunities to sell other courses to customers. When they do see an opportunity, they send a co-worker to this particular customer who specializes in the type of courses the customer is interested in. You set targets and you plan to evaluate the change 6 months from now.

The problem here is that the new process requires a new star schema model. `Sales Manager` is no longer an attribute of the `Customer` dimension. You need to create a new covering fact table that keeps track of which account manager has which courses in their portfolio. There will probably be more changes, such as keeping track of how a bonus should be divided between two account managers when you spot an opportunity and the other scores the final deal. Changing dimensions is especially problematic. Other processes in the organization that have nothing to do with the changes in the sales process might also use those dimensions. Changing the dimensions might mean you have to change queries that feed reports of other departments.

Changes are part of the nature of business intelligence, but pose problems for the same business intelligence. Besides internal changes, as the preceding example describes, you might have changes coming from changed legislation. Or you might think of a new way to use your existing data. Whatever the reason, a data warehouse should be agile. We cannot afford to redesign the data warehouse every time something changes.

As we said, Bill Inmon was the first to criticize the dimensionally modeled data warehouse. His idea, the so-called corporate information factory, has a slightly different architecture than the one shown in *Figure 9.1*. Later, Dan Linstedt proposed a so-called Data Vault table structure. The business intelligence architecture becomes as in *Figure 9.3*:

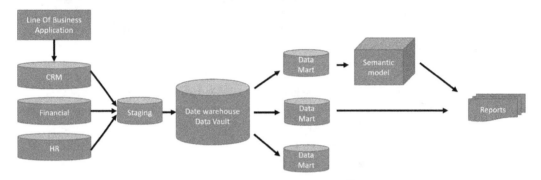

Figure 9.3 – Business intelligence architecture with Data Vault

The general idea is to build a layer between the data's origin, the line-of-business applications, and its analytical use in a dimensionally modeled data mart that is optimized for specific usage. The data warehouse is the layer that should be stable over time. When processes or legislation change, the data itself does not, so the data warehouse should not. In that way, you create a repository to store historical data. The repository, the data warehouse, should be optimized for the long-term storage of historical data. It should be flexible in that when new types of data are introduced, new data should be easy to add to the data warehouse.

By now, we know that when you optimize any type of database for a certain use case, it probably is *not* optimized to do something else. A Data Vault data warehouse is optimized for the long-term, flexible storing of historical data. It will, most of the time, show really bad query performance for the types of queries that we write to create reports. It is not made for using the data at all.

That is why, in *Figure 9.3*, you see Data Marts behind the data warehouse. When you need to create reports and dashboards for **Sales**, you create a dimensionally modeled Data Mart for **Sales**. This Data Mart is a subset of all the data and is loaded using data from the data warehouse. It is optimized for how **Sales** intends to use the data now. When **Sales** decides to reorganize, nothing happens to the data warehouse. We may want to drop the existing

Data Mart and create a new one optimized for the new way in which we work. When **Marketing** wants reports and dashboards as well, we create a separate independent Data Mart for them. That might mean that you have a product dimension in two different Data Marts. The advantage is that both Data Marts are completely independent of each other. Changes in the **Sales** department no longer have an impact on **Marketing**.

> **Note**
>
> A Data Vault data warehouse is optimized for storing data and loading new data, but not for querying data. Its goal is to form a stable basis to build business intelligence solutions.

Dan Linstedt proposed a table structure with so-called Hubs, Links, and Satellites. Hubs, Links, and Satellites are types of tables in the same way as dimension tables and fact tables are types of tables. A database created using Hubs, Links, and Satellites is called a Data Vault. A Data Vault is a hybrid solution that combines normalizing data and dimensional modeling. Let's now have a look at what Hubs, Links, and Satellites are.

Designing Hub tables

As already mentioned, Data Vault distinguishes between three different types of tables: Hub tables, Link tables, and Satellite tables. The first step in the modeling process is to find the Hub tables, or just Hubs. Hubs have a direct relationship to what we called entities in *Chapter 2*, *Entity Analysis*. Data Vault calls these business keys. Each entity has a characteristic that makes it unique. That characteristic was called a logical key in *Chapter 1*, *Introduction to Databases*. In Data Vault, it is called the business key. Each business key (that is, each entity) will be implemented as a Hub table. A sales process will have a Hub for products and a Hub for customers. Product and Customer are the entities involved in a sales transaction.

A Hub table has only four columns. You can see a Hub table example in *Figure 9.4*:

Hub
Primary key
Business key
Load Date
Row source

HubProduct
ProductHashKey
ProductName
LoadDate
Source

Figure 9.4 – Hub table

All the columns in the Hub table are mandatory and you should always create them in order. One of the key things about Data Vault is that we use the same structure over and over again. That means that a Data Vault makes it easy to generate source metadata. Also, **Extract, Transform, Load** (ETL) will be easily generated. The four columns that a Hub table consists of are as follows:

- A hash value of the key
- The business key (can be a combined key made up of several individual columns)
- The date the row was loaded into the Data Vault
- The name of the source system the row originates from

You add these four columns in this order in the table. However, when designing the Hub table, you start by choosing a proper business key. So, let's take a closer look at all four columns, starting with the business key.

Defining the business key

The most difficult part of designing a good, flexible Data Vault is the choice of the business keys to use. This choice leads us back to the discussion on keys in *Chapter 1, Introduction to Databases*. Candidate keys are all columns, or all combinations of columns, that hold unique values. A logical key is a column that holds real value to you and that you need to store that just "happens" to hold unique values. A technical or surrogate key is an extra column with meaningless but unique values that you add to a table for the purpose of having a unique and efficient key. In normalizing data and dimensional modeling, surrogate keys are preferred over logical keys.

Data Vault chooses flexibility over efficiency. Logical keys are preferred over surrogate keys. When you sell products and each product has a unique name, the product name is a good choice for the business key. There are a couple of characteristics that apply to a key:

- A business key consists of one or more columns that are used by the business to search data.
- The values in the column (or columns) you choose to be the key will not change. Preferably, the values *cannot* be changed in the source system.
- The key has the same meaning and grain throughout the whole organization. The value might be formatted differently or might be stored using a different data type in different source systems.

Preferably, a business key is not based on the surrogate key of the source system. When creating a Hub product, the name of the product is a better choice for the business key than the product number when the product number is a surrogate key. One reason for this is that the product number of a given product might change when a new source system is introduced or when you merge with another company. The product itself, and by that we mean the name of the product, doesn't change. But when you are in a situation where, for instance, marketing often changes the name of a product in an effort to boost sales, the product number might be more stable than the name. In that case, the product number would be the better choice to serve as the business key in your Data Vault.

Let's look at an example. Suppose you have a **Customer Relationship Management (CRM)** system with a table called `Relation`. Each relation is assigned a surrogate key in the CRM database when a new relation is entered. This surrogate key is stored in a column called `RelationNumber`. You also have a financial system that handles invoicing and has a table called `Debtor`. A debtor is a customer that still needs to pay an invoice. The `Debtor` table has a surrogate key called `DebtorID`. The generic term *customer* is a synonym of both *debtor* and *relation*.

Since *debtor* and *relation* are the same, the `Relation` and `Debtor` tables will, for the most part, store the same data. However, the `DebtorID` and `RelationNumber` columns that describe the same customer are probably different. Neither `DebtorID` nor `RelationNumber` can be used as the business key.

But suppose in both databases the combination of the name and address is unique. This makes the combination a candidate key. In your Data Vault, you implement a Hub table called `hubCustomer` that has the name and address as its combined business key. The rows in this table will be a union of data from both the `Relation` table from the CRM system and the `Debtor` table from the financial system.

A business key is always unique in the Hub table. Uniqueness is enforced at the database level by implementing a unique constraint.

Implementing a hash key

The business key is not the primary key of the Hub table. You calculate a hash value from the actual value of the business key. This hash value serves as the primary key of the Hub table.

A hash function generates a binary value from an input value in a deterministic way. Deterministic means that the same input always leads to the same output. The same value for the business key always leads to the same binary value. *Figure 9.5* shows an example where the **Peanut Butter** product name is hashed into a binary value:

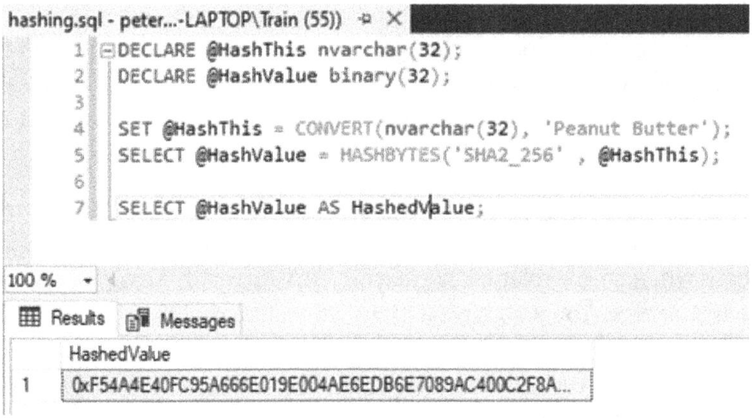

```
hashing.sql - peter...-LAPTOP\Train (55))  ⊅ ×
     1 ⊟DECLARE @HashThis nvarchar(32);
     2   DECLARE @HashValue binary(32);
     3
     4   SET @HashThis = CONVERT(nvarchar(32), 'Peanut Butter');
     5   SELECT @HashValue = HASHBYTES('SHA2_256' , @HashThis);
     6
     7  LSELECT @HashValue AS HashedValue;

100 %   ▾ ◂
   ⊞ Results  ▣ Messages
       HashedValue
    1  0xF54A4E40FC95A666E019E004AE6EDB6E7089AC400C2F8A...
```

Figure 9.5 – Calculating a hash key

By calculating a hash value from the business key and using the hash value, you make sure that the primary key of each Hub table is always a single column. Whether the business key is made up of one column or is a combination of multiple columns, after hashing the business key value, you end up with one value that you can store in one column.

Another important aspect of the hash key is that you can always calculate the hash value from the original business key value. When loading fact tables in a star schema, you need to look up the foreign keys in the dimension tables. These lookups can be expensive. It also means you need to load the dimension tables before you can load the fact table. You need to use the surrogate keys, which might not exist until the dimension tables finish loading. If you can calculate the key value instead of having to use lookups, you can optimize the load process.

The important thing with hashing is that the calculated hash values need to be unique. You can't afford a situation where **Peanut Butter** and **Marmalade** translate into the same hash value. In both Azure SQL Database and Synapse Analytics SQL pools, you can use the HASHBYTES function (as shown in the screenshot in *Figure 9.5*) using different hashing algorithms. It is recommended to use either the SHA2_256 or SHA2_512 algorithm. When you hash a million different product names using the SHA2_256 algorithm, you have a 4.318079×10^{-66} chance of having a duplicate value. When two different input values lead to the same hash value, we speak of a hash collision. A hash value calculated using the SHA2_256 algorithm leads to a hash value that is 256 bits in size. Using the SHA2_512 algorithm doubles the size of the resulting hash value, but also drastically reduces the chance of having hash collisions.

Adding the load date

The date on which the key is entered into the Hub. This date should be generated by the ETL process that loads data into the data warehouse. When you perform nightly loads, that is, you load new data generated during the day into the data warehouse each night, all rows loaded during the night get the same load date. So, we refer to not an exact time that the row is loaded but the date. It is more like a batch code: all rows loaded by the same execution of your ETL process get the same value for `LoadDate`.

You do not want to use a date coming from a source system. With Data Vault, you should always use the same structures. Some source systems will have something such as an inserted date for new rows. Others don't. Your Data Vault needs to be independent of whether the source system provides this functionality. Creating your own load date guarantees that you have a load date and that it uses the same logic to generate it for all rows across the entire Data Vault database.

An extra advantage of using your own dates is that international organizations' rows in the same Hub table may come from source systems in different time zones. That complicates using the dates because what looks like the same date might not actually be the same because it may come from a different time zone.

Adding the name of the source system

Traceability of data is another key aspect of a Data Vault data warehouse. That is why we always add the name of the system where a key was first seen. This sounds simple, but it is not always that simple. A new customer record might have been entered into the CRM database. But on the same day, a row for that new customer is added to the invoicing database as well. During the nightly load of the data warehouse, you see the same business key in two different source systems for the first time. In such scenarios, you use the name of the master dataset if you have master data. Otherwise, you choose one of the systems as being leading and use the name of that system.

Since names are inefficient compared to numbers, you might create a table with a surrogate key and a row for each source system. The Hub table can then store the source system ID instead of the name.

Adding optional columns

The columns described thus far are mandatory and are always used in the order shown in *Figure 9.4*. Optionally, you can add a `LastSeenDate` column and other metadata columns such as a batch ID, an ETL version number, or an ETL package ID.

The `LastSeenDate` column is updated every night during the load process with the current date when a business key is read from the source and is also found in the Hub table. When a business key is no longer read because it was deleted from the source, the `LastSeenDate` column is no longer updated. This makes it easy to trace back which rows are deleted and when that happened.

Designing Link tables

A Link table, or just Link, can be one of the following two things:

- A relationship between two or more Hub tables

- A (business) transaction

Before we go into these two cases, have a look at *Figure 9.6*. *Figure 9.6* shows the column structure of a Link table:

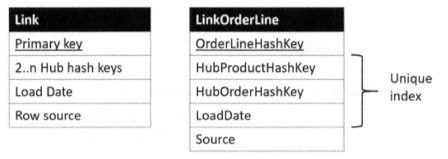

Figure 9.6 – Link table

A Link consists of two or more foreign keys. These can be hash keys from Hubs or from other Links. Like Hub tables, Link tables have a `LoadDate` column and a `Source` column. The primary key of a Link table is the hash value calculated over all the foreign keys together with the load date. The foreign keys are, of course, hash values themselves because they reference the hash keys of the Hub tables. The columns shown in *Figure 9.6* are mandatory and should always be created in the order shown. This makes automatically generating Link tables easier, the same argument as used for Hub tables.

You can choose to add some optional columns, such as `LastSeenDate` and the ETL run ID, as described for Hub tables.

A Link can be a relationship between two or more Hub tables. In *Chapter 2, Entity Analysis*, you learned that there are three types of relationships between entities. The Hubs are, of course, entities, so they have relationships. However, in Data Vault, each relationship is *always* modeled as a many-to-many relationship. The Link table is the bridge table created to support that many-to-many relationship.

The underlying idea of modeling everything as a many-to-many relationship is long-term data warehouse stability. Every relationship can be implemented as a many-to-many relationship because you can limit many to a maximum of one if you want to. A relationship that is one-to-many today can become many-to-many in the future. This is a problem if you modeled the relationship as one-to-many but not if you already modeled it as many-to-many. Your data warehouse is better capable of dealing with a changing world this way. You do pay a price with decreased query performance. But as we have already said, agility and stability over time are more important factors for a Data Vault than query performance.

A Link table can be a relationship between Hubs; it can also be a transaction. Link tables are similar in nature to fact tables in a star schema. They define the grain of the data warehouse. They implement transactions happening in the real process. The example in *Figure 9.6* could be a supermarket where every scanned product leads to a new row in the Link table.

A Link table can reference another Link table. You need this when you need to store facts with different grain. Imagine, for instance, a bank providing loans to companies. Each load is balanced in the books with a risk for the same amount as the loan. A loan of a million is balanced by a risk of a million. The load, however, might be composed of different parts. Each part has a different amount and different conditions. The risk, however, is not split into parts. This would lead to the model shown in *Figure 9.7*:

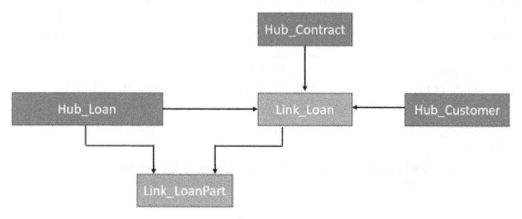

Figure 9.7 – Simple Hub-Link structure

The **Link_Loan** Link table in *Figure 9.7* relates **Hub_Contract** to the **Hub_Customer** Hub table. Even though the bank signs a contract with one customer, we model a Link table in between. The loan itself is composed of parts, where a part by itself is a loan. That means that we need to model a second Link to implement the loan parts.

Because every relationship is modeled as a many-to-many relationship, Hub tables never have foreign keys. A Hub table is basically just a business key uniquely describing an entity that plays a part in the process modeled. For every relationship that entities have, we model a separate Link table consisting of the foreign keys to the entities (Hubs) involved and a load date. This means that a Link table adheres to the following rules:

- A Link consists of at least two foreign keys.

- A Link does not contain any descriptive information.

- A Link is the only table type that implements relationships between entities.

- A Link is used for every relationship, including one-to-many relationships.

- A Link always contains the load data and a reference to a source system.

- A Link should implement the lowest level of detail available from the source systems.

With the Hubs and Links defined, you have the main structure of the Data Vault. But we haven't stored any descriptive data yet. That is where Satellite tables come in.

Designing Satellite tables

Hubs and Links only contain keys. Hubs only contain the business keys of entities, not descriptive data of those entities. Links only contain foreign keys to Hubs or other Links, but again, no descriptive data. We do create the LoadDate and Source columns, but these are not real descriptive data of the entities or the relationships. We add Satellite tables to the main structure of the Data Vault as defined by the Hubs and the Links. Satellite tables, or just Satellites, contain the "real" data, the descriptive data. This, of course, includes non-unique data values such as product color.

A Satellite is a table containing detailed and historical descriptive data columns of a Hub or a Link. *Figure 9.8* shows a Satellite:

Satellite		SatProduct
Hub/link Primary key		HubProductHashSleutel
Load Date		LoadDate
End Date		EndDate
'Actual columns'		Color
Row source		Weight
		Source

Figure 9.8 – Satellite table

A Satellite has the following columns:

- The hash key of the Hub or Link it belongs to
- The load date of the row
- The end date of the row
- The actual descriptive columns that describe the Hub or Link it belongs to
- The source of this descriptive data

Let's take a look at these columns in turn.

The hash key is a foreign key referencing a Hub or a Link. The Satellite as shown in *Figure 9.8* references the `HubProduct` Hub.

The load date is the date when a row is added to the Satellite. Satellites implement what we called **Slowly Changing Dimension (SCD)** type 2 in *Chapter 7, Dimensional Modeling*. All attributes are defined as SCD type 2. Every time a single descriptive value changes, we create a new row to store the new value. The old value remains in the Satellite. This means we can have multiple rows with the same hash key. The hash key together with the load date is unique. They make up the primary key of the Satellite.

The load date is determined by the ETL. You always know when you insert a new row. Depending on the source and the frequency of your ETL, you might not know the exact date (and time) the change occurred. This leads to consistent values throughout the data warehouse and is consistent with what we said about the load date for both Hubs and Links as well.

You may deviate from this approach during the initial load of the Data Vault. You may also deviate from this approach when values change at a higher frequency than your ETL process runs and the source system provides you with all changes and the actual date and time of the change.

The end date of a row is updated when a new row with the same hash key is inserted and the row was the current row for that particular hash key. This is also equal to implementing SCD type 2 in a dimension table. The end date is the only column in a Satellite that can change. As long as a row holds the actual values, the end date is best set to `9999-12-31`. When a newer version of the row arrives, you set the column to the ETL date.

Suppose in the example of *Figure 9.8* that products have unique names that identify the products. The product name was therefore chosen to be the business key. All other characteristics (attributes, columns) are stored in the Satellite or possibly in multiple Satellites. In the example of *Figure 9.8*, we have columns for the product's color and weight.

Lastly, we store the source of the descriptive columns as we do for Hubs and Links.

Adding optional columns to a Satellite

As with Hubs and Links, you can add some optional columns to the previously described mandatory columns. One optional column could be the actual change date of a value next to the data warehouse load date when a source system has that information.

An often-used optional column is a column storing a so-called hash difference. As with implementing SCD type 2, the ETL has to figure out whether an already existing hash key has different values for the columns that we currently find in the source system. That is the trigger to update the current row by making the **End Date** column equal to the current date (this is called to end-date a row) and insert a new row for that hash key. You could check column by column to see whether anything has changed. Or you could calculate a hash value over all the descriptive columns and store this hash value along with the columns themselves. In the ETL, you can calculate the hash as well. When you compare the newly arrived data to the data already in the Satellite, all current rows with the same hash key but a different hash value for the columns have changed. If both the hash key and the hash difference are equal, the row did not change.

Now that we know about the columns in a Satellite, there is one last important remark to make. A Satellite has at least one row for each key in the Hub or Link it belongs to. So, for example, when a new product is introduced in the business, we will add a new key (a new row) to the corresponding Hub table. We then also add a new row to the Satellite with the descriptive columns of that product.

It is now time to look a bit closer at how to design Satellite tables.

Choosing the number of Satellites to use

Have a look at *Figure 9.9*, which shows a simple Data Vault structure:

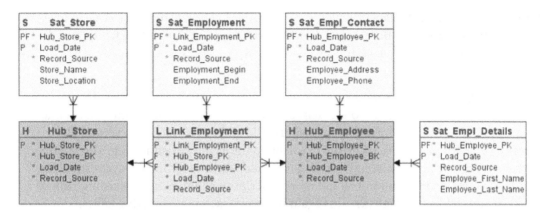

Figure 9.9 – Simple Data Vault model

In *Figure 9.9*, we see two Hub tables, **Hub_Store** and **Hub_Employee**. Because employees work in stores, we have a Link between the two Hubs, **Link_Employment**. The descriptive information describing stores, employees, and employment is stored in Satellites. But notice that the **Hub_Employee** Hub has two Satellites, and not just one. Employee address information is stored in the **Sat_Empl_Contact** Satellite, whereas personal information is stored in the **Sat_Empl_Details** Satellite. We could have stored all information in one Satellite, but chose to spread the columns over two Satellites. You can create as many Satellites per Hub or per Link as you want. This choice is what makes Data Vault such a flexible model when it comes to storing historical data.

Suppose that in the first iteration of implementing a data warehouse, you only needed the employee detail information, that is, the first name and last name of the employees if we stick to the example of *Figure 9.9*. You design the **Hub_Employee** Hub and the **Sat_Empl_Details** Satellite as in *Figure 9.9*. You also implement the ETL to load both tables in nightly batches and you start using the data by creating reports. A year later, people using the data warehouse need employee contact information, so you decide to add this information to your Data Vault. You have two options to accomplish this task. Either you alter the existing **Sat_Empl_Details** Satellite or you add a second Satellite.

Altering an existing table can be a challenging task. At first glance, it might be as simple as running an ALTER TABLE statement against the table to add the columns. But you then need to change the existing ETL code as well to start loading the newly added columns. *And* you need to test whether all queries using your Satellite still work. In other words, you need to do an impact analysis of your intended change and you might find that you have to do a lot of work to keep everything working correctly after the change.

Or you create a new Satellite table. Of course, you'll need to write new ETL code to load data into the table as well. But because you change nothing in the existing tables, you do not have to change existing ETL processes and other queries using the already existing tables.

An advantage of Data Vault modeling is that you can start small. The smallest Data Vault database has one Hub table and one Satellite to go with the Hub. Every time you extend your Data Vault with new information, you just add tables. You can add a second Hub table and add a Link table to relate the new Hub to the already existing Hub. The existing Hub does not change when doing this. You then add Satellites for the newly created Hub and Link tables. When you want to add more descriptive columns for an existing Hub or Link, you just add a new Satellite. This again has no impact on the existing tables.

There are actually two important advantages described here. One is that you can start small and incrementally implement a data warehouse. This goes very well together with agile development and working according to Scrum principles. The second advantage is that you can always extend the functionality of a Data Vault with the upfront guarantee that it has no impact on already existing functionality.

More arguments may lead to a design with multiple Satellites per Hub or Link than an incremental approach to implementing your Data Vault. These arguments are as follows:

- Divide columns over multiple Satellites based on the type of information they contain.
- Divide columns over multiple Satellites based on the rate of change.
- Divide columns over multiple Satellites based on origin.

Dividing columns over multiple Satellites based on the type of information they contain

One option is to look at the meaning of the columns you need to store. The example of *Figure 9.9* makes a distinction between contact information and personal information. If these types of information are often used separately, you might store them separately.

This argument focuses mainly on the usage of the data. That is a bit strange because Data Vault is not about how to use data. You will probably create a dimensionally modeled Data Mart that you will use to serve the data to users.

Dividing columns over multiple Satellites based on the rate of change

A Satellite stores the entire history of a Hub or Link key. Each time a single column gets a new value, a new row is added to the Satellite. Suppose that for some reason, employees in the example of *Figure 9.9* get new phone numbers very often. Their names, however, might change once or twice, when they either marry or get divorced, but are otherwise stable. If you store all columns in a single Satellite, you need to store the name again and again every time the phone number changes.

You could create a Satellite for columns that have values that (almost) never change. You could then design a second Satellite to store the columns that change regularly. This Satellite is now much smaller than the combined Satellite would have been. This makes the impact on storage of storing all the history smaller. And that, of course, means lower storage costs for your Azure database or Synapse SQL pool that you use to implement the Data Vault.

Dividing columns over multiple Satellites based on origin

The third option is to look at the origin of the data. You may have multiple source systems that deliver different columns about the same business key. This is a very natural thing to do in Data Vault. Just look at the Source column that you add to all tables. This already suggests that you create separate tables when you have separate sources. What would you enter in the Source column when half of the columns in the Satellite have their origin in the CRM system and the other half come from the financial system? Making two Satellites seems logical and solves the problem. The advantage is that the traceability of the data to where the data originated drastically improves. That is also a key design principle of Data Vault and a prerequisite in lots of real-world scenarios.

This last strategy, using Satellites based on the source system, is the preferred way. But factoring in cost should always be a consideration.

Before we move on to explain all the theory by using an example, we need to spend some time understanding the use of hash keys.

Using hash keys

Hash keys were introduced in Data Vault 2.0 and play a central role in the design. One advantage of using hash keys is that both Hubs and Links can (and often will) have a composite key. This makes the keys large and inefficient. By creating a single hash value, the key becomes more efficient. This is not true for Hubs with small keys. In that case, hash keys are likely to be more inefficient. But there are other arguments. One other argument is that we want all tables to have the same structure. So, we also use a hash key, even when that is less efficient.

The most important advantage of using hash keys is the efficiency gain of the load process. To explain this, consider a star schema with surrogate keys. You need to load the dimension tables first. The surrogate keys are created during the insert of the new dimension rows. After all the dimensions have finished loading, you can start to load the fact table. The ETL process, which gets fact rows with source keys for the dimensions, needs to look up the surrogate key in the dimension table based on the source key it has. This process has two disadvantages. The first disadvantage is that you have to wait on loading the fact table until all the dimension tables have finished loading. The second disadvantage is that looking up all the surrogate keys in the dimension tables takes a lot of time.

You would get the same disadvantages when working with surrogate keys in Data Vault. Your load would look something like the upper schematic of *Figure 9.10*:

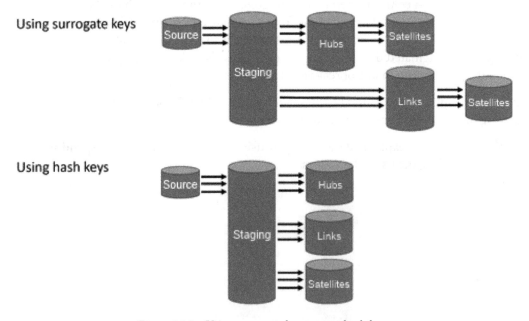

Figure 9.10 – Using surrogate keys versus hash keys

The big advantage of hash keys is that they are deterministic. Each time you calculate the hash key, you can be sure you're getting the same value. Suppose you use unique product names to be the business key of the **HubProduct** Hub. When you load a Link table referencing the **HubProduct** Hub, you do *not* need to look up the key. You just calculate the hash key based on the product name. The same is true when entering rows in the **SatProduct** Satellite. You do *not* need to look up the key. You just calculate the hash key again.

You can load the Hubs, Links, and Satellites independently because you calculate the hash key for each of these tables. That provides you with the opportunity to load the data warehouse according to the second schematic shown in *Figure 9.10*. This load strategy is easier to set up because you have fewer dependencies. You can start loading all the data warehouse tables as soon as all the data is in staging. You can load the tables in a random order. You can also load as many tables in parallel as your system will allow you to. Together, this can significantly reduce data warehouse load times.

An important consequence is that you cannot implement foreign keys in Data Vault. A foreign key would demand the referenced hash key to be present in the referenced table. But we are already used to this. Azure Synapse Analytics SQL pools do not support foreign keys anyway.

Now that we know about the table types to use, we need to bring it all together into an overall Data Vault design. The next section uses an example to do just that.

Designing a Data Vault structure

Let's start with the modeling process. There are two ways to design a Data Vault, both using the same general approach. This approach consists of the following steps:

1. Identify the business keys, the natural entities involved in the process you try to model. This creates the Hub tables.

2. Identify the relationships between the created Hubs. This creates the Link tables.

3. Identify the descriptive information to store in the data warehouse. Create Satellites to store this data.

4. Possibly re-group the descriptive columns based on type, rate of change, or source system and create Satellite tables accordingly.

There are two starting points you can choose to start modeling a Data Vault structure. These are as follows:

- Translate a normalized database design into Hubs, Links, and Satellites.
- Determine the entities by analyzing the business processes to model.

The easiest way is to take an existing normalized source database as the starting point. You investigate each table to see whether it has a logical, natural key. If so, it is most likely a Hub. When the natural key is a combination of attributes, it is more likely a Link. For it to be a Link, the individual elements of the key need to be unique somewhere. They actually need to be foreign keys to Hub tables. A natural key can be a composite key of elements that are, by themselves, not unique, but are unique when combined. Such a key is most likely the key of a Hub table.

Have a look at the database schema shown in *Figure 9.11*. This is a slightly simplified version of the exercise you made in *Chapter 2, Entity Analysis*:

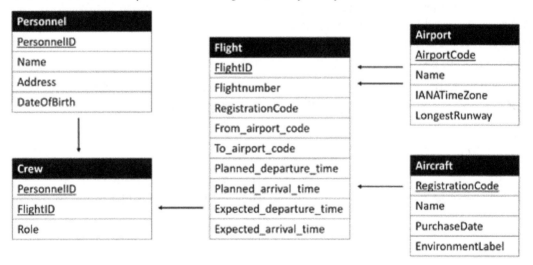

Figure 9.11 – ERD flight planning system

Let's translate this ERD into a Data Vault structure.

Choosing the Hubs

The first step in our modeling effort is to find the business keys. The primary key of a table can be a business key, although it often turns out to be a surrogate key. You can also look for unique constraints or unique indexes on tables when you have a real database you work with. Columns that are enforced by the database that store unique values can very well be business keys.

Using the ERD of *Figure 9.11*, we find the following Hub tables:

- **Hub_Airport**: The name of the airport is the business key.
  The `Airport` entity also has an `AirportCode` column. This could be a surrogate key, in which case we do not want to use it as the business key. But the **International Air Transport Association (IATA)** gives every airport a unique three-letter code. Amsterdam airport, for example, has the letters SPL. If this is meant by `AirportCode`, it would be a perfect business key. So, on second thought, `AirportCode` will be used as the business key.

- **Hub_Aircraft**: Each aircraft has a unique registration, like each car has a license plate. Assuming this registration is meant by the `RegistrationCode` column, this code is the business key.

- **Hub_Personnel**: The `Personnel` table is a candidate to become a Hub table. The business key to choose is less clear. Theoretically, the `Name` column is the business key because we use the name to address a person. Unfortunately, the name is not unique. Even when we combine the `Name`, `Address`, and `DateOfBirth` columns, we may not have unique values. That means that there is no alternative but to use the `PersonnelID` surrogate key as the business key.

- **Hub_Flight**: This is a tough one. Is `Flight` a Hub table or a Link table? Both seem to be possible. It is not always immediately clear. You could look at how the information is going to be used in reports. In this case, we choose to implement `Flight` as a Hub table. Read on for the explanation.

There are two important factors to consider when determining whether a table is a Hub table or a Link table. These factors are as follows:

- Can the entity exist on its own?
- Should the entity be stored separately?

Transaction tables are tables storing entities that cannot exist by themselves. They hold information about what occurred while executing business processes. A bank transaction cannot exist without two bank accounts where money is transferred between them. The bank account itself, on the other hand, can exist without transactions transferring money in or out of the account. `Transaction` will become a Link table, whereas `Account` becomes a Hub table.

The second criterion, whether the entity is stored separately, can be explained by looking at the `Order` entity. The `OrderID` column could be used as a business key for a Hub table. But the `Order` table is a transaction-oriented table that mainly consists of foreign keys. This suggests it should become a Link table. In a sales transaction in a store where a product is exchanged for money, there is no separate "thing" such as an order. The transaction becomes a Link. When an initial quote is turned into a purchase order with a purchase order number that you provide when sending an invoice, the order is an actual entity and becomes a Hub table even though the end result is that a product is exchanged for money. You may need to create a list of all purchase orders created in a certain period. They need to be stored.

Back to our example. We created a Hub table using the `Flightnumber` column as the business key. This entity is the service (product) an airliner offers to its customers. The website of the airliner shows a list of all flights to be executed in the upcoming months in order for you to book a seat on a specific flight. Each row holds information on a planned flight that we intend to execute, like in a `Product` table that holds information on products that we intend to sell.

When you look closer at the original table, you see columns for `FlightID` and `Flightnumber`. These are different. A flight number is usually not unique. An airliner may operate a daily flight from the same airport to the same airport at the same time of day. Each time we actually fly, the same flight number is used, but a different flight ID is issued.

The `Flight` table is not normalized properly. The flight number identifies the `From_airport_code`, `To_airport_code`, `Planned_departure_time`, and `Planned_arrival_time` columns. This should be a separate table. This is the table we turn into a Hub table. The flight ID identifies the `Flightnumber`, `RegistrationCode`, `Expected_departure_time`, and `Expected_arrival_time` columns. This table will be turned into a Link table.

All this leads to the Hub tables shown in *Figure 9.12*:

Hub_Airport	Hub_Aircraft	Hub_Personnel	Hub_Flight
AirportCodeHash	RegistrationCodeHash	PersonnellDHash	FlighttnumberHash
AirportCode	Registration	PersonnellD	Flighttnumber
LoadDate	LoadDate	LoadDate	LoadDate
Source	Source	Source	Source

Figure 9.12 – The Hub tables

The next step is to find the Links.

Choosing the Links

When searching for Link tables, you look for transactional tables or tables that represent relationships between entities. Tables that were not turned into Hub tables in the previous step are also likely candidates to be Link tables. In our example, this leads to the following Link tables:

- **Link_Flight**: This table represents the actual occurrence of an airplane flying on a specific date to a specific airport. You can see the `Hub_Flight` table as a service. The `Link_Flight` table is the actual delivery of that service. This is where the aircraft, airport, and flight number come together.

- **Link_Crew**: There is a many-to-many relationship between the `Hub_Personnel` and `Link_Flight` tables because the crew on an airplane consists of multiple people.

The discovered Link tables are shown in *Figure 9.13*:

Link_Flight
Link_FlightHash
To_airport_codeHash
From_airport_codeHash
RegistrationCodeHash
FlightnumberHash
FlightDate
LoadDate
Source

Link_Crew
Link_CrewHash
PersonnelIDHash
Link_FlightHash
LoadDate
Source

Figure 9.13 – The Link tables

Notice the **FlightDate** column in the **Link_Flight** table. We need this to make a unique key. It is the main difference between the flight number and flight ID.

The last step is to design the Satellites.

Choosing the Satellites

In our example, we add all the data from the source system into the data warehouse. Columns from the source system that we chose to be the business keys are already part of the Hubs. All the other columns need to be stored in Satellite tables that have a relationship with either a Hub or a Link table. Using the same arguments that we know from entity analysis and normalizing data, this leads to the Satellites shown in *Figure 9.14*:

Sat_Airport
AirportCodeHash
LoadDate
EndDate
Name
IANATimeZone
LongestRunway
Source

Sat_Aircraft
RegistrationCodeHash
LoadDate
EndDate
Name
PurchaseDate
EnvironmentLabel
Source

Sat_Personnel
PersonnelIDHash
LoadDate
EndDate
Name
Address
DateOfBirth
Source

Sat_Hub_Flight
FlightNumberHash
LoadDate
EndDate
Planned_departure_time
Planned_arrival_time
Source

Sat_Link_Flight
Link_FlightHash
LoadDate
EndDate
Expected_departure_time
Expected_arrival_time
Source

Sat_Crew
Link_CrewHash
LoadDate
EndDate
Role
Source

Figure 9.14 – The Satellite tables

As explained in the *Choosing the Hubs* section, planned and expected departure and arrival times belong to different tables.

When we add all the tables we created together into an ERD, we get the ERD shown in *Figure 9.15*:

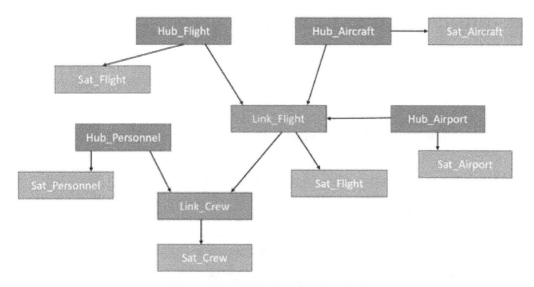

Figure 9.15 – Flight planning Data Vault

As we said at the start of this section, there are two ways to start designing your Data Vault. The first way is to take your normalized source database and decide table by table whether the table should become a Hub table or a Link table. The drawback of this approach is that you might focus on a single source too much. You want to create a single Data Vault to store all your (relational) data no matter which source system it is coming from. Some people do not consider focusing on a source database as a liability. You could create a Data Vault structure per source system.

Part of the advantage of creating a data warehouse is to integrate all your data into a single database. To reach that goal, you should do an entity analysis independent of the source databases. Or you could start from a "master" system. This is a system that is appointed by the business as leading or as the most important system. After implementing a Data Vault based on this system, you can start adding data from other sources by mapping entities from those new sources to existing Hubs. You can then add new Satellites to the Hubs you mapped your new entities to. You might have to create some new Hub and Link tables as you go along when you can't map your entities to existing Hubs.

When integrating multiple sources into a single Data Vault, finding suitable business keys is the biggest challenge.

You now know how to design a Data Vault structure. We will end this chapter by looking at some features to make a Data Vault more usable.

Designing business vaults

The first version of the Data Vault theory introduced the notion of Hubs, Links, and Satellites. This provided the flexibility over time that we needed for data warehousing. Data Vault 2.0 has a couple of adaptions to the first version. An important change was the introduction of hash keys. But Data Vault 2.0 also introduced the concept of a business vault. This addresses the fact that the structure of a Data Vault as discussed until now will rapidly become complex and difficult to query. The latter leads inevitably to bad query performance.

The introduction of the business vault led to two other new terms: Raw Data Vault and Operational Data Vault. Both terms mean the same and refer to a "pure" Data Vault structure with Hubs, Links, and Satellites as discussed so far in this chapter. The Raw Data Vault stores raw data coming directly from the source database. The columns are stored in different table structures, but the data itself is not changed in any way.

In *Chapter 7, Dimensional Modeling*, we discussed many reasons why you need data warehousing to do reporting and analytics. A Data Vault can integrate multiple sources and stores data historically. The data quality, however, is the same as in the source databases. Also, as just mentioned, the query performance is probably worse than when querying directly on a source database. This is why the architecture of *Figure 9.3* has Data Marts behind the Data Vault. The Data Mart does deliver good query performance, plus you can cleanse the data in between the raw Data Vault and the Data Mart to improve data quality.

The danger of this approach is that some data cleansing rules might have to be repeated for multiple, possibly all, Data Marts. Some data-specific business rules apply throughout a company, independent of process or department. Applying rules just once in a central location is better than repeatedly performing the same actions. This is where a business vault comes in. A business vault is a Data Vault where generic business rules are applied to the data to increase the usability of the Data Vault. One cleansing transformation you can think of would, for instance, be to standardize the data. You can think of the following:

- Formatting phone numbers by preceding them with a country code, such as +31, and deleting braces that may or may not surround the area code

- Translating US state codes to full state names

- Removing synonyms such as Holland and the Netherlands or USA and the United States to always use a standardized consistent name

- Adding missing address information based on ZIP codes

The mentioned transformations are just some examples of general rules that may apply to your data. The important part is that you do not want to prepare your data for a specific use case. That is what a Data Mart is for. But you also don't want to repeat the same transformations for each Data Mart individually.

You could create a separate Satellite to store the cleansed data. The Hub has a Satellite with raw data that can be analyzed when you need to know, for instance, what type of data issues you have in source databases. By using the same Hub table, it becomes easy to compare the raw data to the cleansed data.

You can also deduplicate data. In that case, you would create a new Hub table. You would also create a new Link table that links the Hub table from the Raw Vault to the newly created Hub table that only stores the deduplicated data. This Link makes it easy to see which rows have been combined to form a new row in the deduplicated table.

You could also just create a new Link table. This so-called Same-As Link has two foreign keys, both referencing the same Hub table. One foreign key references the "real" row, the other the duplicate row.

Consider as an example a situation where you have customers in a business-to-business scenario in a Hub table with `CompanyName` as the business key. You find three rows with the `TrainSQL`, `TrainSQL Inc.`, and `Train'SQL` key values. As you decide that TrainSQL is the real name of the customer, you create a (`TrainSQL`, `TrainSQL Inc.`) row and a (`TrainSQL`, `Train'SQL`) row in a newly created Link table. The first value references the real name, the second the duplicate you found. You would, of course, store the hash keys of the names, not the real names as we did here, to clarify how a Same-As table works.

The idea is that we keep the raw data, but also have a cleansed version available. When you create a Data Mart, you can choose which version of the data is a better fit for your use case based on the use case you create the Data Mart for. A business vault is not a completely new layer in the architecture of *Figure 9.3*. It is not a new and separate database. The term *business vault* just specifies that we enriched the raw data to increase the consistency and usability of the data in the Data Vault.

There are more structures you can add to your raw Data Vault besides adding Same-As Links and Satellites with cleansed data. Let's look at a couple of them.

Adding a Meta Mart

When building a data warehouse, you will find you have a lot of metadata. Metadata describes the actual data. Names and data types of columns are an example of metadata.

You could create a database that stores the metadata. To start with, you want a list of all databases that are used as the source database for the Data Vault. This table can be used to provide a value for the `Source` column that is mandatory in all your Hubs, Links, and Satellites.

You also might want to track which columns in the data warehouse are based on which source columns. Column names might be different. The business vault might even contain calculated columns. The expression used for the calculated column is metadata and needs to be known for traceability and data validation purposes.

Other interesting metadata is metadata related to the security aspects of your data. Is your data sensitive and who may or may not see the data? What do you do when someone executes their right to be forgotten?

Adding a Metrics Vault

In most cases, you need to measure the performance of your ETL process. You need to monitor the ETL process to proactively maintain the process and to have data to use when the need to troubleshoot problems arises.

The type of information you can think of is as follows:

- Throughput. How many GBs of data can you process per hour?
- The start and end times of the overall ETL process and per table:

  a. Load time of staging

  b. Processing time of the data validation process

  c. Load time of the Raw Data Vault

- The number of rows processed:

  a. The number of new rows added per table

  b. The number of changed rows

  c. The number of invalidated rows

- The wait times of processes waiting for other processes to finish first.

The Metrics Vault can be a Data Vault structure by itself. A dimensional Metrics Vault is also fine and increases usability.

Adding an Error Mart

When implementing a data warehouse or a data lake, you need to validate the data. You can create an Error Mart that keeps track of all rows that didn't pass validation, including the reason they failed validation. This information can then be used to improve an automated validation process or as a feedback loop to the organization.

Using Point-in-Time tables

A Point-in-Time Satellite, also called a PIT table, is an optional part of the business vault. It helps in retrieving data from a Hub with multiple Satellites. In such a case, you may have to retrieve rows from different Satellites but make sure that the values you read were active during the same period. Each Satellite having its own beginning and end date might make this difficult.

PIT tables combine the load dates from the different Satellites. *Figure 9.16* shows an example of a PIT table belonging to the **HubProduct** Hub:

Hub_Product_Hash	PIT_Load_Date	Price_Laad_Date	Properties_Load_Date
158B332NPQR56X	1-jan-2020	NULL	1-jan-2020
158B332NPQR56X	2-jan-2020	2-jan-2020	1-jan-2020
158B332NPQR56X	3-jan-2020	3-jan-2020	3-jan-2020

Figure 9.16 – Product PIT table

The key of the PIT table is the combination of the hash key of the Hub table and the load date of the PIT table itself. The example of *Figure 9.16* assumes that the HubProduct Hub has two Satellites, one with pricing information and another with all other product properties.

On January 1, 2020, a key was loaded in the Satellite with product properties but no row existed with pricing details. On January 2, a new row with pricing details was entered. That means we also entered a new row in the PIT table. On January 3, both Satellites received a new row with updated price and property information.

When you add a new row to the PIT table on every day of the calendar, it becomes easy to retrieve the correct values for a Hub key for a specific day. PIT tables are created to improve the query performance of selecting time-related data.

Adding Bridge tables

Bridge tables, like PIT tables, are part of the business vault. They also have the improvement of query performance as their sole goal. They simplify queries by bringing relevant keys together:

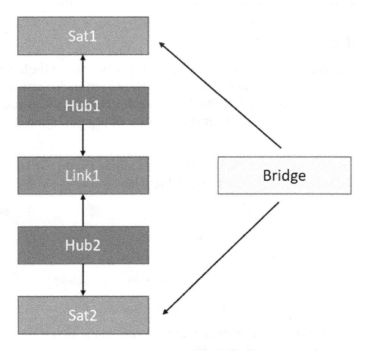

Figure 9.17 – Bridge table

Have a look at *Figure 9.17*. When you write a query that needs a column from the **Sat1** table and you need to combine it with a column from the **Sat2** table, you need to join five tables together. The Bridge tables store keys of the **Sat1** table in combination with the key of the **Sat2** table. That means that when using the Bridge table, you only need to join three tables. When you have more complex examples, the difference between using a Bridge table or joining the original tables will be even bigger.

Adding a hierarchical link

When modeling a Hub table, such as **HubEmployee**, you might find that one employee reports to another employee. You need a Link table to implement this relationship. This is basically the same as an entity that has a relationship with itself. Because Data Vault always uses many-to-many relationships, the Link table will look like the one shown in *Figure 9.18*:

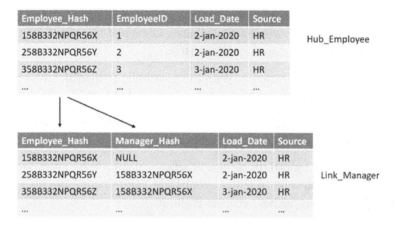

Figure 9.18 – Hierarchical Link

The **Link_Manager** Link table contains a foreign key to the row of an "ordinary" employee. It also contains a foreign key to the row of the manager of this "ordinary" employee.

PIT tables and Bridge tables help to query the complex structure that a Data Vault quickly becomes. Hierarchical links are a normal modeling technique used for entities that have a relationship with themselves.

Implementing a Data Vault

So far in this book, design chapters were followed by provision and implement chapters. A Data Vault data warehouse, however, is a relational database. You can implement a Data Vault using either a SQL database or a Synapse Analytics dedicated SQL pool. The choice is largely economical. The total size of the Data Vault will matter. You need to estimate the total size. Once you know the tables and the data types of the columns, you can calculate the size by estimating the number of rows. You then need to know something about the usage and the performance you need. Size and performance together determine whether a SQL database will do the job or whether you need a Synapse SQL pool. In both cases, you can use the skills learned in earlier chapters to implement the Data Vault.

Summary

Data Vault modeling is a mix between normalizing data and dimensional modeling. It is designed to provide a flexible way to store detailed, historical data. By using Hubs, Links, and Satellites, you create a database in which you never need to alter an existing table. All changes can be handled by adding new tables. That makes a Data Vault more stable over time than other database designs.

Data Vault is also very standardized. This enables tools to automatically generate Data Vault structures from existing normalized databases. On top of that, the ETL process that loads the tables can also be generated for the most part.

As a third benefit, Data Vault is scalable. With the possibility to load all tables in parallel, it can leverage powerful hardware and load vast amounts of data in short periods of time.

Using the business vault functionalities improves the usability of data. However, using Data Marts behind the Data Vault is still a best practice.

A Data Vault optimizes the long-term storage of relational data. When you need to deal with other types of data, you want to replace your Data Vault or combine your Data Vault with a data lake. That is what the next chapter is about.

Exercise

Figure 9.19 shows the ERD of the Northwind database once again. Translate the normalized database into a Data Vault using Hubs, Links, and Satellites:

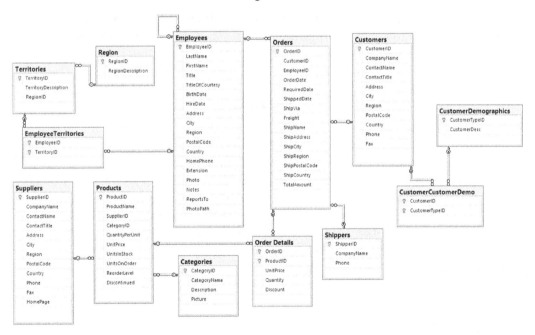

Figure 9.19 – Northwind

You can find the ERD shown in *Figure 9.19* in the downloads for this book for improved readability.

10

Designing and Implementing a Data Lake Using Azure Storage

In *Chapter 9*, *Data Vault Modeling*, you learned how to design a Data Vault data warehouse. It is a flexible and scalable data warehouse. The flexibility refers to the fact that it is agile: it can adapt easily to different circumstances, such as new source databases or other reporting requirements. It is, however, a relational database implementation. Relational databases are good at handling structured data.

What if you have data in JSON documents stored in Cosmos DB? What will you do with web logs, error logs, and other semi- or non-structured data that is also interesting to analyze? Are you going to transform that data into table structures?

Instead of combining with a Data Vault data warehouse, you might implement a data lake. This chapter teaches you the why and the how of designing and implementing a data lake. You will learn about the following topics:

- Background of data lakes
- Modeling a data lake
- Using different file formats
- Provisioning an Azure storage account

Technical requirements

You will need the following if you wish to successfully complete this chapter:

- You will need a connection to the internet and a modern browser.
- You will need an Azure subscription with permission to create new services.

Background of data lakes

Let's start with the definition of a data lake:

A data lake is an environment where you collect and store (vast amounts of) raw data in its original format.

The term *data lake* comes from a water analogy. You can make money using water, just as you can make money by using data. But you will need to store the water somewhere until you find a use case for it. You don't necessarily know beforehand what that use case is going to be. This means you need a cheap and easy way to store the water. Putting all your water in bottles is optimal when you want to sell it as drinking water. But pouring water out of a bottle over a house that is on fire in the hope of extinguishing the fire would probably be useless. So, you wouldn't bottle it until you started selling drinking water:

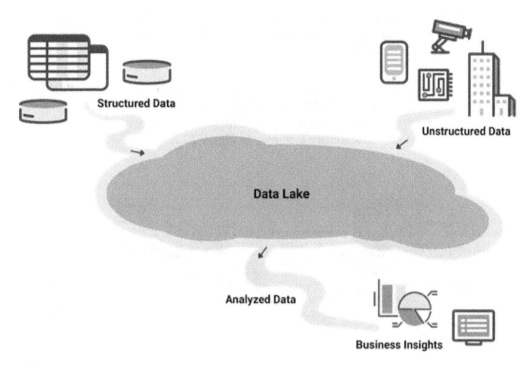

Figure 10.1 – Data lake

Data is analogous to water. When you store it for later use, but you don't necessarily know all the (possible) use cases (yet), you don't change it into a particular format. Why store the data in a textual log file in tables when you might find out later that a data scientist wants to analyze the raw data in the file? Only import data into a dimensionally modeled data mart when you know the reporting demands that you want to use the data for, just as you'd only bottle water when you want to sell it as drinking water.

Until the time of actual use, you want to keep the data in its original format in a large reservoir of data. Reservoir and data lake have become synonyms in this regard in recent years. Because we gather vast amounts of data nowadays, a data lake implementation should be scalable. Azure storage accounts provide you with the scalable storage needed to store limitless amounts of raw data.

A data lake should be capable of storing all types of data. This includes structured data as well as semi-structured and unstructured data. The goal is to save all kinds of data for later use. The idea is comparable with a data vault. A data vault is optimized to store structured data for later use. When you have your use case, you take a subset of the data and you create a dimensionally modeled data mart for it. Data lakes extend this concept with all types of data instead of limiting us to structured data. As with a data vault, once you have a use case, you transform a subset of the data to optimize it for the intended usage (if necessary). That leads to an architecture as shown in *Figure 10.2*:

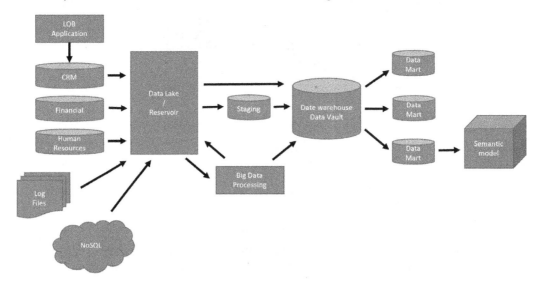

Figure 10.2 – Modern data warehouse architecture

There are multiple variations that you can derive from *Figure 10.2*. One implementation that is often used is to export all data from source databases into files. These files can be stored in the data lake. The data lake becomes the data warehouse where we keep all historical data. As we have already seen in *Chapter 8, Provisioning and Implementing an Azure Synapse SQL Pool*, you can implement a star schema Synapse SQL pool based on the files you have stored in the data lake. In this scenario, you do not need the data vault as shown in *Figure 10.1*. The staging layer, as depicted in *Figure 10.1*, is not a separate physical layer but is made up of staging tables inside Synapse Analytics.

Another variation takes the definition of *data lake* more literally. Why export data originating in tables to files when you might import that data into tables again when loading your data mart? The data lake concept might comprise a data vault to store your data that is structured in nature and origin. Alongside the data vault, you use an Azure storage account to store all other data in files. The storage account and the data vault together form the data lake.

Having all your historical data stored as files in a storage account and storing that same data in a data vault, as *Figure 10.2* suggests, does not make a lot of sense. But a storage account might be a landing zone that you upload upload the on-premises data to an on-premises SQL pool. It sits there only temporarily until you load the data into the next stage: the data warehouse. This constitutes a more classic BI architecture.

Synapse Analytics can also be used for data virtualization purposes. In that case, the data warehouse in *Figure 10.2* has external tables that reference files and folders in the data lake. The data marts are physical implementations of databases created in Azure SQL Database or Synapse Analytics that use the external tables to retrieve the data from the data lake. Each data mart can be a different physical database or just a logical grouping of tables implemented in a single database.

The main concept of a data lake is to not perform too much processing on data before you actually know what you want to do with the data. Storing data in a binary format in a data lake is doing just that. A data lake is basically a big hard drive that you use to store bits and bytes. The hard drive has no clue what is in the files or what the bits and bytes stand for. When we think of a use case, we then transform the data to optimize it for the use case.

Figure 10.2 shows a rectangle with the text **Big Data Processing**. A data lake is just storage. When you need to process the data, you need compute power. Storage and compute are separated. This allows you to implement the processing of the data in the data lake using the language and platform that suits your use case. With a SQL database, storage and compute are not separated. You store the data in the database and use T-SQL to work with the data. Although this is perfectly fine in most scenarios, you need more flexibility when you do not know beforehand what type of data you have and what sort of processing you want to perform on that data.

Now that you know why you may need a data lake, let's look at how you set up a data lake.

Modeling a data lake

A data lake is, in essence, nothing more than a limitless hard drive that we use to store files. Everyone knows that if you do not carefully consider a folder structure to use on your personal computer, you will end up with a mess and it will become almost impossible to find your files. It is no different for your data lake. Although we cannot speak of data modeling when designing a data lake, creating a structure is really important for a successful implementation. That is even more true when the data in the data lake is made available to end users, such as data analysts working with Power BI or data scientists working with Python notebooks.

When creating a folder structure, you need to take the following things into account:

- The source of the data
- The functional meaning of the data
- Security: Who may or may not need to have access to the data?
- Who and which applications will use the data?
- Is the data stored permanently or temporarily?
- Is the data loaded in batches or in real time?

Before we look at a folder structure, however, we need to look at data lake zones first.

Defining data lake zones

The first step in setting up a data lake is to divide the data lake into zones. Zones can be implemented as top-level folders. But depending on the usage, you can also create separate containers or even separate storage accounts for individual zones. You'll learn about provisioning storage accounts and creating containers in the *Creating a data lake filesystem* section. Possible zones are as follows:

- **Raw data**: This is where data arrives. You could also call it the landing zone or possibly staging. This zone is close to the original concept of a data lake in that the data is not processed or transformed in any way. It is a direct copy of the data at the source.

- **Processed data/standardized data**: The first step to perform after the data is uploaded to the raw zone is to validate the data. There might have been issues during the export, or compression formats might not be recognized. This will impair further processing of the data. There can also be functional issues with the data that we might want to report back to the organization.

  When there are exports coming from multiple systems with different settings, we might have issues such as dates being formatted as dd-MM-yyyy in some files, but formatted as MM-dd-yyyy in other files. You could standardize these types of issues to improve the usability of the data. You do not want to correct data. We still store raw data, including errors and data inconsistencies. But you can validate and standardize the data.

- **Transformed data/curated data**: This zone can be compared to the business vault. Some business rules apply to all data and all use cases. You can apply these business rules to your data each time you use it, or do it once and use this curated data.

You can add additional zones to your data lake. An example of how you can divide your data lake into zones is shown in *Figure 10.3*:

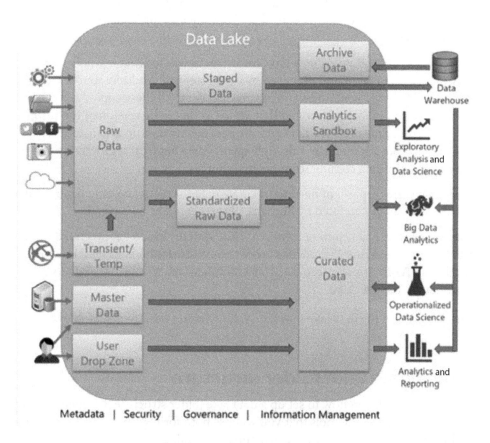

Figure 10.3 – Zones in a data lake

We recognize the three zones already mentioned. We also see a couple of other zones in *Figure 10.3*. Let's have a look at them:

- **Staged data**: When you decide to use the data lake for data not coming from relational databases and implement a relational data warehouse on the side, the raw data zone can be used as a landing zone. All relational data, from on-premises databases, cloud databases, and **SAAS** (Software as a Service) applications, is copied or uploaded to the raw data zone. From there, it is transformed as needed and put into staging. The data warehouse reads the prepared data from staging.

- **Archive data**: Storing data in an Azure storage account can be a lot cheaper than storing it in a SQL database or Synapse SQL pool. You might export unused data from the data warehouse back to a storage account to save money.

- **Analytics sandbox**: Data analysts or data scientists might request data from the data lake. Depending on the data they want to use, security restrictions may apply. You can create a separate zone for such a request. You can grant access to this sandbox to only those users who may use the data. You can also, for instance, anonymize the data in the sandbox.

- **Master data**: Organizations may have master data. Master data is where a central copy of a dataset is kept and managed by a data steward. Multiple source systems may have tables and containers storing product data. All these copies vary slightly. Therefore, a central copy is created that serves as the single version of the truth. This is often already cleansed data and the best quality data an organization has. You want to take advantage of this high-quality data but treat it differently to unmanaged raw data coming from normal sources.

- **User drop zone**: You might need to integrate data that is not coming from database systems but that is generated by humans. Targets could be an example.

- **Transient/temp**: Data coming from real-time systems has to be integrated into the data lake as well. Azure Event Hubs can write events streamed in real time to the cloud to an Azure storage account. This event data can then be integrated with the other data in the data lake.

Once you have divided the data lake into zones, those zones need to get a folder structure. So, let's look at that.

Defining a data lake folder structure

There is no definite guide on how to set up a data lake folder structure. It depends on the data you need to store and the tools that you will use both for ingestion and processing the data. We do, however, need some folder structure to organize the data in the data lake.

The first step is to decide how the folder structure and zones work together. One option is to create a folder structure within a zone. A zone is then a folder at the highest level. This option is implemented most often and has the ETL process and security as the main arguments to do so. Often, users should only be able to access data from the curated zone. Having the zone as the top level of a hierarchy makes it easy to provide permissions at this level. The top-level folders are in sync with the ETL steps where data "flows" through the data lake with each preparation step we perform on the data. Within the zone, you can create functional areas such as sales data, web log data, and telemetry data. You can see an example in *Figure 10.4*:

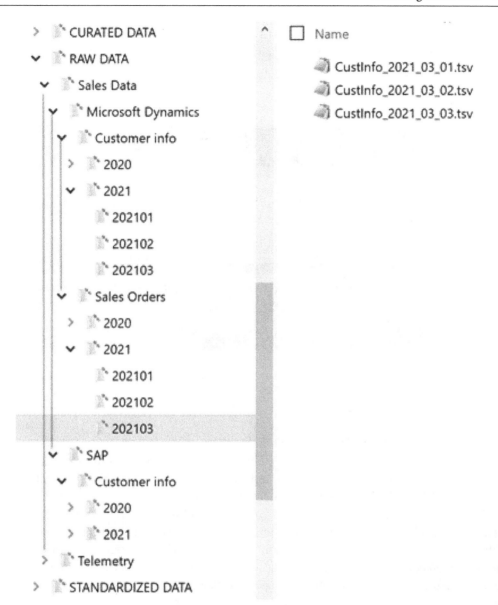

Figure 10.4 – Data lake folder structure

You could, as an alternative, create functional top-level folders. You could create a top-level folder called Sales Data. Within this folder, you could create subfolders for raw data, standardized data, and curated data. This approach focuses on the functional aspect, on the meaning of the data. When specific subsets of your data need to be treated differently, for example, from a GDPR perspective, this approach can make sense.

When using Azure Data Lake Storage Gen2 to implement your data lake, you need to map your zones and folders to Azure storage account functionality. *Figure 10.5* shows a high-level diagram of how a storage account is set up:

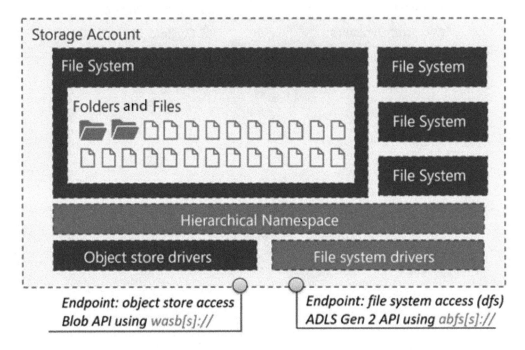

Figure 10.5 – Storage account data lake components

The highest level is the storage account itself. You can create as many storage accounts as you need in an Azure subscription. You might want to use separate storage accounts to implement **DTAP (Development, Test, Acceptation, Production)** environments. An important aspect of a storage account is that you specify at the account level whether you want a hierarchical namespace. Choosing a hierarchical namespace turns your storage account into Data Lake Storage Gen2, as depicted in *Figure 10.5*. Without the hierarchical namespace enabled, you create a Blob storage account. Blob storage is cheaper than Data Lake Storage. You may want to implement the archive data zone as Blob storage to save money.

Data Lake Storage has two main advantages over Blob storage. The first is that it provides better performance. The second is that you get folder-level security. Once you have a storage account provisioned, you create containers to store data in. A Blob storage container is a flat namespace, which basically means that you don't have "real" folders. That means that you cannot set access permission at the folder level. Users can access the entire container or nothing at all.

The hierarchical namespace setting provides you with the ability to create folders within containers. Containers are then called filesystems. Filesystems allow access control at the folder level. As you can see in *Figure 10.5*, you can create multiple filesystems within a single hierarchical namespace.

In *Figure 10.4*, we assumed you have a hierarchical namespace and implement your data lake in a single filesystem. In some scenarios, using multiple filesystems might be beneficial.

Let's go back to *Figure 10.4*. You can see that the **RAW DATA** folder contains a **Sales Data** folder and a **Telemetry** folder. Data coming from different functional areas within the business is put in different folders. The sales data itself is sourced from two different systems: Microsoft Dynamics and SAP. Different folders are created for the different source systems. The folder structure here is just a suggestion. You could have created the **Microsoft Dynamics** and **SAP** folders directly underneath the **RAW DATA** folder. In many situations, you might have just a single source system per functional area, in which case the extra **Sales Data** folder doesn't add anything but complexity.

In *Figure 10.4*, we get customer data and sales order data from Microsoft Dynamics. Since this data is different, we put it in a different folder. We further divided the folders into time slices, with folders for years and subfolders for months. Again, you could have started with year folders on a higher level and stored data from all systems from the same period together in the same folder. However, since we rarely set security based on time but often based on the functional area, it makes setting up security a lot easier to follow the suggested structure of *Figure 10.4*.

Let's look a bit closer into time slicing your data.

Designing time slices

The first question you should ask when creating time-based folders as in *Figure 10.4* is which time to use. You can use the time of ingestion in the data lake or the time associated with the data in the source system. This is like the load date in the data vault where the default is to use the actual load date as calculated by the ETL process and not a transaction date.

In a data lake, data can be pushed into the data lake or pulled into the data lake. Pushing data in is when you have an Azure Event Hubs or a Stream Analytics job writing events to the data lake. The dates in the folder structure will typically be the actual ingestion date. Using Stream Analytics, you set up partitioning in Stream Analytics. You can specify, for instance, a path pattern to be `{datetime:yyyy}/{datetime:MM}/`. Stream Analytics will automatically create the folder structure as depicted in *Figure 10.4*.

Data can also be added to the data lake using batch ETL processes. This is a pull method, specifying we "pull" data from source systems into the data lake. This pulled data can be divided into two separate use cases: the data can be transactional in nature or you can implement snapshot data.

In *Figure 10.4*, you can see a folder called **Customer info**. This data can be meant to show a complete picture of all the customer information at a certain time. In the example, the customer info data is stored in full in the files shown. This can be an easy solution for small reference datasets. This is called snapshot data.

Figure 10.4 also shows a folder called **Sales Orders**. You will most likely extract data in increments based on the transaction date. When you read data a couple of days after the fact, you want the transaction date to be leading and not the load date. Data like this is append-only data in most cases.

After thinking about which timestamps to use, the second design decision to make is to decide at which grain you create folders. Do you create a folder per year with all files in that folder? Or do you create a subfolder per month in the year folders? And do you create a folder per day within the month folders or not?

The tools you use and the usage patterns you envision must be leading. That is a bit strange since you want the data lake to just be storage and the tools you use can be chosen at a later date when you have new use cases for your data. However, this argument is especially valid for the curated data zone. When implementing an Azure data lake, you will most likely choose a tool to do the processing needed between raw and standardized data and between standardized and curated data. So, take this tool into account when setting up a folder structure. PolyBase, for instance, can read all files from a single folder as a single external table. Files in subfolders are not included. Reading an entire month's worth of data becomes easy when all files are in a single month folder. Implementing daily increments into a Synapse SQL pool might be easier when you have a folder per day.

New use cases also mean new usage patterns that you may not have foreseen. This is, in most cases, not relevant for the raw and standardized zones. These are most likely not made available to users anyway. It is, however, relevant for the curated zone.

Other tools may have other demands on file and folder names.

The last thing to mention about the folder structure you choose is the files you store in them. Only store files with the same structure in the same folder. As was just said for time slicing, you often read all files in a folder as a single external table or linked dataset. This leads to errors when you store CSV files with different structures in the same folder.

Storing JSON files from the same Cosmos DB container in a folder is fine even when the schema differs per file. This is hard to avoid and the processing logic should handle the complexity anyway.

To finalize this section, we come back to the different zones in the data lake. Different zones will have a different folder structure. The structure as shown in *Figure 10.4* might be a good fit for the raw data zone and possibly for the standardized data zone. In the curated data zone, it will make more sense to leave the structure based on the source system. Storing the data by functional area makes more sense. For analytics, it most likely doesn't matter where the data comes from. You just want all the relevant data to be stored together so it is easy to find and use.

Now that we have learned about folder structures to use in a data lake, we need to focus on the file types to use.

Using different file formats

Storage is said to be cheap nowadays. That does not mean that we should waste our money. When you store a lot of data and pay for the volume of data stored, it pays to compress your data.

When you use the on-demand options of Azure Databricks or Azure Synapse Analytics to process data in a data lake, it also pays to reduce the total duration of the processing. Both storage and processing are arguments to have a look at the different big data file formats that come from the Hadoop platform.

PolyBase in Synapse Analytics can work with delimited text files, ORC files, and Parquet files. Azure Data Factory can also work with AVRO files. Other processing platforms might even have other file types. AVRO, Parquet, and ORC evolved in Hadoop to decrease the cost of storage and compute. With the right file format, you can do the following:

- Increase read performance
- Increase write performance
- Split files to get more parallelism
- Add support for changing schemas
- Have better compression ratios

So, let's have a look at these file formats.

AVRO file format

AVRO files store data and metadata together. The schema of the data, in other words, the columns stored in the actual data, is stored in JSON format. AVRO can handle schemas that change over time, called schema evolution, where, for instance, new columns are added to newly added files in existing folders.

The data in itself is stored in a row-by-row fashion in a binary format. This binary format is compact and efficient. It is also flexible in the data structures it allows. You can, for instance, use arrays within rows. Languages such as C#, Java, and Python can deserialize AVRO. That means you can use Azure Synapse Spark pools to process data in these files. You can alternatively use Azure Data Factory to read AVRO files.

AVRO can be used in the earlier zones in your data lake. When you process the data as a whole, for instance, when validating all the data, the row-based format is efficient. Also, AVRO files have a better write performance than Parquet and ORC files. And lastly, the capability of handling schema changes at this level makes the data lake more agile.

Parquet file format

Parquet files store data in a columnar way. This is comparable to the columnstore index that Synapse SQL pools use. When you query a small subset of columns in a dataset, it is far more efficient to store the data column by column instead of row by row. When storing the data row by row, the entire row needs to be read and then parsed into the columns you need. With a column-by-column storage mode, you just retrieve the columns you need. An extra advantage is that compression works better on columnar data because it is easier to find similarities in the data to compress.

Parquet files also provide the possibility to use nested data structures as AVRO does.

Parquet files can be used more downstream in the data lake. Data scientists or data analysts might query files in the curated zone of the data lake. When they issue OLAP-style queries, Parquet files may give them the best performance.

ORC file format

ORC stands for **Optimized Row Columnar**. ORC files use a columnar way to store data just as Parquet does. It reaches high compression ratios. It uses indexes and metadata to optimize queries.

Parquet and ORC files are optimized for read access. Parquet has a more flexible nature. When you need to store nested data structures, you want to use Parquet. ORC, on the other hand, compresses better and is more capable of predicate pushdown. That means that when you add a filter to a query, the data can be filtered at the file level instead of reading in an entire file before filtering the data.

On top of choosing the proper file format, you want to compress your data to save on cost. Multiple compression techniques can be used. In Synapse, you specify the used compression method in the external file format along with the specification of what file format is used. For details, browse to `https://docs.microsoft.com/en-us/sql/t-sql/statements/create-external-file-format-transact-sql?view=sql-server-ver15&tabs=delimited`.

Now that we know a little about file formats, let's have a look at file size.

Choosing the proper file size

Azure Data Lake Storage Gen2 and the compute engines from Spark (Databricks and Synapse Analytics Spark pools) as well as Data Factory are optimized to perform better on larger file sizes. When queries need to work with a lot of small files, the extra overhead can quickly become a performance bottleneck.

Apart from performance considerations, reading a 16 MB file is cheaper than reading four 4 MB files. Reading the first block of a file incurs more costs than reading subsequent blocks.

Optimal file sizes are between 64 MB and 1 GB per file. This can be a challenge in the RAW zone where you may not have a lot of control over file sizes.

Good testing to find the optimum number of files and file sizes for the compute you use is key here.

After learning the basics of the setup of a data lake, let's start implementing one by provisioning an Azure storage account.

Provisioning an Azure storage account

The first step in setting up a data lake is to create a storage account:

1. Open the Azure portal at `portal.azure.com` and log in.

2. Open the menu in the Azure portal and click on **Create a resource**.

3. In the search box, type `storage account` and press *Enter*.

4. Click on the **Storage Account** tile.

5. Click on **Create**.

    The **Basics** tab of the **Create storage account** page opens as shown in *Figure 10.6*:

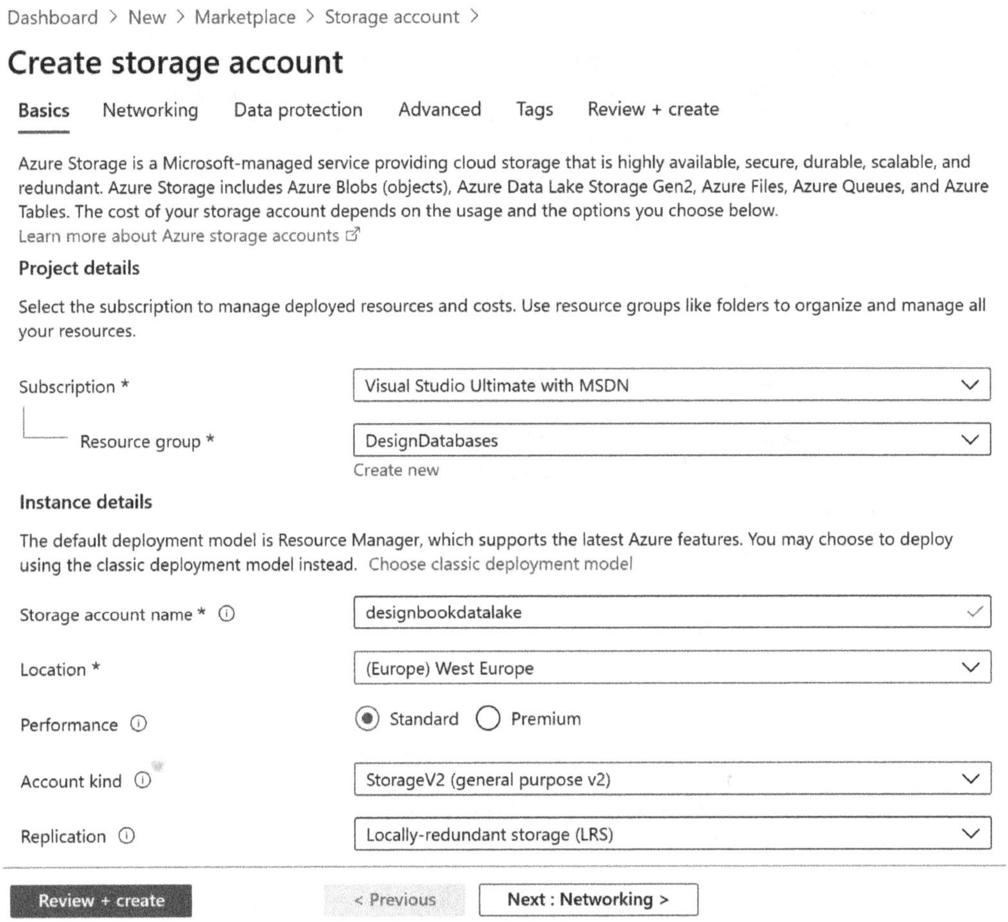

Dashboard > New > Marketplace > Storage account >

Create storage account

Basics   Networking   Data protection   Advanced   Tags   Review + create

Azure Storage is a Microsoft-managed service providing cloud storage that is highly available, secure, durable, scalable, and redundant. Azure Storage includes Azure Blobs (objects), Azure Data Lake Storage Gen2, Azure Files, Azure Queues, and Azure Tables. The cost of your storage account depends on the usage and the options you choose below.
Learn more about Azure storage accounts ☐

Project details

Select the subscription to manage deployed resources and costs. Use resource groups like folders to organize and manage all your resources.

Subscription *	Visual Studio Ultimate with MSDN ⌄
└─ Resource group *	DesignDatabases ⌄
	Create new

Instance details

The default deployment model is Resource Manager, which supports the latest Azure features. You may choose to deploy using the classic deployment model instead. Choose classic deployment model

Storage account name * ⓘ	designbookdatalake ✓
Location *	(Europe) West Europe ⌄
Performance ⓘ	⦿ Standard ◯ Premium
Account kind ⓘ	StorageV2 (general purpose v2) ⌄
Replication ⓘ	Locally-redundant storage (LRS) ⌄

[Review + create]   < Previous   [Next : Networking >]

Figure 10.6 – Create storage account Basics tab

The initial settings are similar to the earlier services you created. The storage account name should be unique worldwide. It can contain only lowercase letters and numbers and can be a maximum of 24 characters in length.

1. Select your subscription to use.

2. Select the `DesignDatabases` resource group that you created in earlier chapters.

3. Name your storage account `designbookdatalake`, appended with your initials to make it unique.

4. Created the other resources from the previous step in this book. In my case, that is West Europe.

The first setting that is specific to an Azure storage account is the **Performance** setting. You choose the **Premium** option when you need a storage account to store virtual machine disks, that is, VHD files. In all other cases, you leave the default option of **Standard** selected.

The **Account kind** setting doesn't leave you with a lot of options. For backward compatibility, you can still choose the **General-purpose V1** option and the **BlobStorage** option. However, it is recommended to use the default option of **General-purpose V2**.

The **Replication** setting is more interesting to consider carefully. The options to choose from are as follows:

- **Locally redundant storage (LRS)**
- **Zone-redundant storage (ZRS)**
- **Geo-redundant storage (GRS)**
- **Geo-zone-redundant storage (GZRS)**
- **Read-access geo-redundant storage (RA_GRS)**
- **Read-access geo-zone-redundant storage (RA_GZRS)**

Locally redundant storage (LRS)

LRS replicates your data three times within a single data center in the primary region. This means that for every file you store, Azure creates three copies and stores the copies on different disks. When a server or disk within Azure fails, you still have two copies left. Azure will detect the failure and automatically create a third copy again and store it on still-operational hardware. This ensures that you do not lose data when hardware in Azure fails. You do not see or notice anything about LRS. To a user, it is like you just have a single copy of the file.

LRS is the cheapest option. When you can restore your stored files yourself in any way after files are lost despite the redundancy of Azure, this is a good option. Or, when your data may not leave the country you are in, LRS may be the option you need to choose.

LRS does not protect against the failure of an entire data center due to, for instance, power outages or disasters such as flooding. You might opt for ZRS to mitigate that risk.

Zone-redundant storage (ZRS)

An Azure region (in the settings called **Location**) is divided into Availability Zones. Availability Zones are physically separate locations within the region. An Availability Zone is made up of one or more data centers equipped with independent power, cooling, and networking. A data center can be a physical building where all the actual hardware is located.

Because data centers are physically separated, a failure in one data center might leave another data center untouched. A power outage in one data center does not mean the other data center also has no power.

With ZRS, Azure will again create three copies of each file you store in your storage account. This time, however, they make sure that all three copies reside in different data centers belonging to different Availability Zones. You are not only protected against a server or hard drive failure, but even against failures of an entire data center. Your data does not leave your primary region. When you choose, for instance, West Europe to be the location of the storage account, data remains within West Europe.

The cost of storing data goes up when choosing ZRS. The availability goes up as well. But what if the entire region of West Europe is unavailable? All your zones are in the same region. When you need to protect against an entire region being unavailable, you need to choose GRS.

Geo-redundant storage (GRS) and geo-zone-redundant storage (GZRS)

GRS will make not three but six copies of each file you store. Three copies are stored in the primary region, the one you specify with the **Location** setting. Three additional copies are replicated to a paired region. You basically have LRS storage but both in the primary region as well as in the paired region.

The data in the secondary region is not available for you to use with GRS. When the primary region becomes unavailable, you can perform a failover to the secondary region. The secondary region then becomes the primary region and your data becomes available again.

Using GRS, you have LRS in each region. That means that you need to do a failover to the secondary location in case of a power outage in the data center of the primary region. You can avoid that with GZRS.

GZRS is like GRS but the primary region is configured as ZRS instead of LRS.

Read-access geo-redundant storage (RA_GRS) and read-access geo-zone-redundant storage (RA_GZRS)

You can choose to make copies of your files available for use in the secondary region by choosing the read-access option. You can configure an application to use the secondary region whenever you want. You do not have to wait for failover to complete.

In Azure, you pay for what you get. Storing files in a storage account becomes more expensive when you choose an option that provides more uptime for your data. For exact **service-level agreements** (**SLAs**) and pricing, we refer you to the official documentation at `https://azure.microsoft.com/en-us/pricing/details/storage/`. You can switch a storage account from one type of replication to any other type after it is created.

Let's go back to configuring the storage account (please note that these steps are a continuation from the steps you set previously):

1. Keep the default option (**Standard**) for the **Performance** setting.

2. Also, keep the default option (**StorageV2 (general purpose v2)**) for the **Account type** setting.

3. Choose the **Locally-redundant storage (LRS)** option for the **Replication** setting.

4. Click on **Next: Networking**.

   After clicking on **Next: Networking**, you get to the **Networking** tab. This tab is about connectivity only. Later, we will set access permissions. We'll now define from where we can reach our storage account. If you choose the **Private endpoint** option, you have to select a private endpoint to use or you create a new private endpoint. A private endpoint is a private IP address belonging to an Azure **Virtual Network** (**VNet**). Only services that are also part of this network can connect to the storage account.

   You can also select one of the public endpoint options. In that case, a public IP address is used for the storage account. You can still limit connectivity to certain Azure VNets you create.

   Private endpoints are arguably the best choice here from a security perspective. But you might have users or applications that are not (and possibly cannot) be part of your VNet. In that case, a public endpoint is unavoidable.

5. For now, keep **Public endpoint (all networks)** selected.

    The second part of the **Networking** tab is about routing. The **Microsoft network routing (default)** option is the default and recommended setting for almost all use cases. This setting means that when a client from outside of Azure wants to connect to the storage account, their request is routed to the closest Azure entry point. This means that when you connect from, for example, Australia to a storage account in Europe, the request is routed to an Australian Azure data center and then sent onward over the Microsoft network. This network is meant to deliver premium network performance with high reliability.

    When you select the **Internet routing** option, the same request is routed over the public internet to Europe to enter Azure in Europe. This can save money on Azure networking costs but may or may not be as fast and as reliable.

6. Keep the default setting of **Microsoft network routing (default)** selected for the **Routing preference** setting.

7. Click on **Next: Data protection**.

    After clicking on **Next: Data protection**, you get to the **Data protection** tab. This tab is about the ability to restore data that you inadvertently delete or change.

    Turn on point-in-time restore for containers enables you to restore an entire container to a previous state. This option is only available for general-purpose v2 storage accounts. You cannot use this option for Azure Data Lake Storage Gen2, an option we will get to on the next tab. In testing scenarios, this option can be very helpful, as well as when you need protection against users or applications changing or deleting blobs.

    When turning on **Turn on point-in-time restore for containers,** you also need to turn on **Turn on soft delete for blobs**, **Turn on versioning for blobs**, and **Turn on change feed**. Since we cannot combine any of these settings with Azure Data Lake Storage Gen2, we will not discuss these properties further.

8. Click on **Next: Advanced** to move to the next tab.

    The **Advanced** tab starts with a couple of security settings, as you can see in *Figure 10.7*:

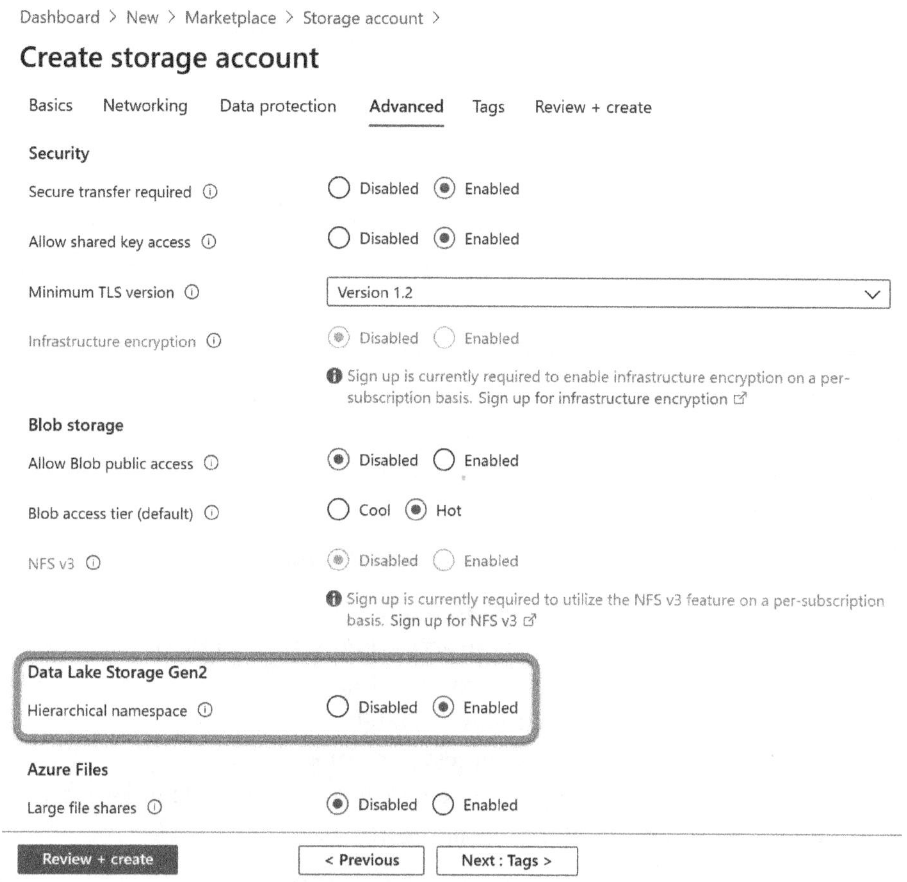

Dashboard > New > Marketplace > Storage account >

Create storage account

Basics   Networking   Data protection   **Advanced**   Tags   Review + create

Security

Secure transfer required ⓘ   ○ Disabled ◉ Enabled

Allow shared key access ⓘ   ○ Disabled ◉ Enabled

Minimum TLS version ⓘ | Version 1.2   ⌄ |

Infrastructure encryption ⓘ   ◉ Disabled ○ Enabled

⚠ Sign up is currently required to enable infrastructure encryption on a per-subscription basis. Sign up for infrastructure encryption ☐

Blob storage

Allow Blob public access ⓘ   ◉ Disabled ○ Enabled

Blob access tier (default) ⓘ   ○ Cool ◉ Hot

NFS v3 ⓘ   ◉ Disabled ○ Enabled

⚠ Sign up is currently required to utilize the NFS v3 feature on a per-subscription basis. Sign up for NFS v3 ☐

Data Lake Storage Gen2

Hierarchical namespace ⓘ   ○ Disabled ◉ Enabled

Azure Files

Large file shares ⓘ   ◉ Disabled ○ Enabled

[Review + create]   [< Previous]   [Next : Tags >]

Figure 10.7 – Create storage account Advanced tab

Unless you have a very good reason to *not* want **Secure transfer required** to be enabled, you keep this setting enabled. The same goes for the third setting, **Minimum TLS version**, which you keep at the default unless you have a reason not to. You may want to put a bit more thought into the **Allow shared key access** setting.

There are three main ways to get access to a storage account: Azure **role-based access control (RBAC)**, using a shared key, or using **shared access signatures (SASes)**. You implemented RBAC already in *Chapter 8, Provisioning and Implementing an Azure Synapse SQL Pool*. You created an Azure Active Directory application and added that application to the **Storage Blob Data Contributor** role in the data lake account belonging to the Synapse workspace. You then created a database-scoped credential to provide the Synapse SQL pool access to the storage account. This approach is the recommended best practice.

Sometimes it is easier (or even necessary) to use keys to access the files. A shared access key (also called a secret key) provides full access to a storage account. Only in trusted situations, for instance, when used by a trusted in-house application, is using a shared access key an option.

SASes, on the other hand, provide granular control over what action (such as read or write) can be done on which service during which time period from which IP address. SAS can be created for user delegation, in which case it is backed by an Azure Active Directory account.

You can enforce the best practice by setting **Allow shared access key** to **Disabled**.

In a scenario where customers can, for example, download instruction videos for products they bought from your storage account, you may want to allow anonymous public read access to blobs and containers. When implementing a data lake, this is a security risk.

9. Change the **Allow Blob public access** setting to **Disable**.

    The next setting is the **Blob access tier (default)** setting, which you can set to either **Cool** or **Hot**. Each blob you store has an access tier of **Hot**, **Cool**, or **Archive**. **Hot** is best used for files you use frequently. **Cool** is better (read: cheaper) for infrequently used files, and **Archive** is meant for files you rarely use. It can take hours to read a file that is stored using the **Archive** access tier. For each file you upload, you can specify the access tier. The setting you find here is merely the default for when you do not specify the access tier when uploading new files.

10. Leave **Blob access tier (default)** at the default setting of **Hot**.

11. Change the **Hierarchical namespace** setting to **Enable**.

    This last setting transforms a "normal" storage account into Data Lake Storage Gen2. This optimizes the files for big data analytics workloads, in other words, for Spark notebooks, Data Factory data flows, and Synapse Analytics. You can access files from all of these services when using normal blob containers. That would be cheaper but provides you with less performance when doing analytics on the data.

    Maybe even more important, using Data Lake Storage Gen2 provides the ability for both Azure RBAC and POSIX-like **access control lists** (**ACLs**). The name **Hierarchical namespace** refers to the fact that we can provide access control at the file and folder levels instead of only at the container level. The last is the case for a regular storage account that has a flat namespace (**Hierarchical namespace** is disabled).

Once you have a storage account, you have to tell Azure how to use your storage account. When storing files in a flat namespace, you have to create a container. A container is called a filesystem when you have **Hierarchical namespace** enabled. Another difference is the URL you use to connect to the storage account. You use the `wasbs` prefix to connect to a container when creating an external data source in Synapse, as follows: `wasbs://<ContainerName>@<StorageAccountName>.blob.core.windows.net`. When using Data Lake Storage Gen 2, you have to use an `abfss` prefix instead, such as `abfss://<ContainerName>@<StorageAccountName>.blob.core.windows.net`.

The last two options are irrelevant when creating a Data Lake storage account.

12. Click on **Next: Tags**.

13. Click on **Next: Review + create**.

14. Click on **Create**.

15. Wait for the storage account to be provisioned and then click on **Go to resource**.

You just created a storage account. An account is nothing yet. It is just a container. The next step is to actually start using it by creating a filesystem.

Creating a data lake filesystem

A storage account is a container resource for multiple different services that you can implement. The first step is to tell the storage account what you want to create. The four services that are part of a storage account are as follows:

- Queues
- Tables
- File shares
- Containers

You can see a tile for each of these options on the **Overview** blade as shown in *Figure 10.8*:

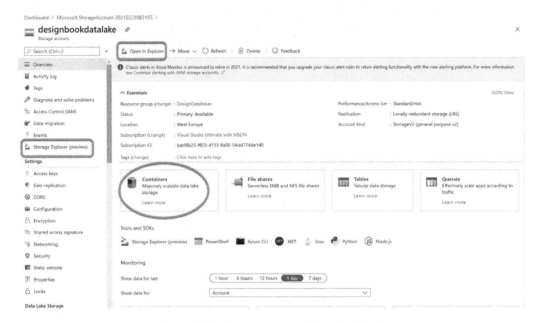

Figure 10.8 – Storage account Overview blade

Queues enable application developers to set up the asynchronous processing of events by having an application put messages in a queue and having another system process these messages. The **Tables** feature is a NoSQL key-value database. File shares are used when you have multiple Azure virtual machines that need access to the same files using the **Server Message Blocks** (**SMB**) protocol. When implementing a data lake, you need to create one or more containers. Here's the fourth option:

1. When on the **Storage Account Overview** blade, as shown in *Figure 10.8*, click on the **Containers** tile.

2. Click on **+ Container** at the top of the **Containers** page.

3. Enter northwind for **Name**.

4. Click on **Create**.

    You just created a container. When using Data Lake Storage Gen2, this is also called a filesystem. Inside a filesystem, you can have folders and files.

5. Click on **northwind**, the name of the container you just created.

6. Click on **+ Add Directory**.

7. Enter Raw for **Name** and then click on **Save**.

8.   Also create a new folder named `Curated`.

9.   Click on the **Raw** folder.

10. Click on **Upload**.

11. Select the **dimCustomer_data.csv** file from the downloads of *Chapter 8, Provisioning and Implementing an Azure Synapse SQL Pool*.

12. Click on **Advanced** to open the advanced properties.

There are a couple of options to consider, although the defaults will normally do. You can choose from three different blob types: **Block**, **Append**, and **Page blobs**. Page blobs are used when storing virtual machine VHD files. Append blobs can best be used when you append new data to the file regularly, like in a log file. In data lake scenarios, we often upload complete files such as CSV or Parquet files. In the case of the CSV file we are currently uploading, the **Block blob** option is the best.

We already discussed the **Access tier** option when provisioning the storage account. When provisioning a storage account, you merely set the default. At this point, you choose the actual access tier for this file. The default should, in most cases, be the right one.

A blob is internally broken down into blocks. Block sizes may vary but are limited to a maximum of 50,000 blocks and vary in size from 64 KB to 100 MB. The best size to choose depends on your workload and bandwidth. You should keep the default of 4 MB. You might test with smaller block sizes to increase upload times when working with limited bandwidth. You might gain read performance improvements on big files when the block size is increased.

Even though you clicked **Upload** from a certain directory, you can specify a subfolder to upload your file to.

Data in your storage account is encrypted. You can have your entire container encrypted using the same key. Or you can create separate encryption scopes to use different keys for different files. Keep the default **Use existing default container scope** setting to use the default container encryption key.

13. Keep all the default settings under the **Advanced** option.

14. Click on **Upload**.

You can work with the files in your storage account from the Azure portal or by using Azure Storage Explorer. When on the **Home** blade of the storage account, you can find Storage Explorer in the left-hand side menu of the portal. We already introduced Azure Storage Explorer in the *Using PolyBase to load data* section of *Chapter 8, Provisioning and Implementing an Azure Synapse SQL Pool.* We'll leave it up to you to connect to the data lake you just created.

You just created a single storage account. But is that enough or do you need multiple accounts? Let's discuss some arguments about whether to use one or multiple storage accounts.

Creating multiple storage accounts

You can provision as many storage accounts as you like. There are four high-level considerations when it comes to choosing how many storage accounts you need:

- DTAP
- Data diversity
- Cost sensitivity
- Management overhead

Let's look at each of them in a little more detail.

Considering DTAP

DTAP stands for **development, testing, acceptance, and production**. Each stands for a separate environment used in a separate stage in the creation of, and working with, a data lake. You probably need to create some code to move data from the source systems into the data lake and then through the different data lake zones. This is a work in progress that should be separated from business processes working with data already in the data lake. You will most likely want to have small test datasets to create the logic and workflow to save development time.

After new software is written, it must be tested. You may have separate datasets to perform all sorts of tests.

After testing is complete, you let the business see what you created. It can provide feedback that may have to be implemented before the business can actually start working with the data.

There will most likely be different demands for the storage account for the different stages or environments as described. You may opt for GZRS for your production data. But do you want to spend money on this kind of availability for your other environments as well? Can you permit developers to work on the same virtual network as where the real data lives?

Apart from these more technical arguments, there is the question of who should pay for storing the data. Maybe different departments should be billed for development and production.

When all the settings as described in the *Provisioning an Azure storage account* section are the same for all four environments *and* the bill should go to the same subscription, you may implement DTAP as top-level folders in a single storage account. But most likely, you'll want separate storage accounts for each environment.

Considering data diversity

When looking back at *Figure 10.2, Zones in a Data Lake*, you can see all the different zones that you might have. The *Defining Data Lake zones* section described the different use cases for the different zones.

As said when discussing DTAP, you specify a couple of settings in the storage account setting. Using a single storage account and creating folders or containers for all the zones leaves you with the same settings for all zones. The **Replication** setting especially should probably be different for different zones.

The settings you choose when setting up your storage account should be determined by the expected usage and the required SLAs while minimizing the cost. Different accounts for different zones may make a lot of sense from a cost perspective.

Different subsets of your data may also be subject to different security demands. Maybe you want data to stay in a certain region. Or maybe you want data to be completely separated from the rest in a separate network. In both cases, you will need separate storage accounts.

Considering cost sensitivity

As we have seen throughout this book, Azure can be cost-effective, but only when you design for it. An important argument in the previous two sections is that some settings make it more expensive to store your data. Especially with a data lake that is expected to be large in size, you should choose the cheapest possible option for your intended use case. With a lot of different datasets in the data lake, that has to be considered on a more detailed level than the data lake as a whole.

Another important factor here is billing. When using multiple storage accounts, you can create them in separate subscriptions for separate billing. Or you use the same subscription but you use the tagging feature to clarify on the bill which cost belongs to which department or project.

Considering management overhead

The main drawback of multiple accounts is that you need to create them and, more importantly, maintain them. Everything gets slightly more complicated. IT personnel need more time to oversee the architecture, set up security and alerts, and keep the system healthy. In general, more resources need more attention, which costs money as well.

Now that we can make a well-informed choice on how many storage accounts to create, we are ready to implement a data lake in Azure.

Summary

A data lake plays a central role in the modern data warehouse architecture. In this chapter, you learned what a data lake is. You learned what zones a data lake may consist of and how to set up a folder structure for those zones. You also learned about some different file formats to use.

You implemented a data lake by provisioning an Azure storage account.

The next chapter will give you a high-level insight into how to copy data from source systems into a data lake and how to propagate the data further downstream.

Section 3 – ETL with Azure Data Factory

Data lakes, data warehouses, data marts, and potentially semantic models need to be loaded with new data regularly. This process is called the **Extract, Transform, Load (ETL)** process. This section is about getting data out of the source systems you designed in *Section 1* into the analytics systems you designed in *Section 2*.

This section comprises the following chapter:

- *Chapter 11, Implementing ETL Using Azure Data Factory*

11
Implementing ETL Using Azure Data Factory

In *Section 1* of this book, you learned how to design and implement operational databases using Azure SQL Database and Azure Cosmos DB. Operational databases are databases where new data comes into existence; for instance, data pertaining to new customers and data regarding new orders is added to the database.

In *Section 2* of this book, you learned how to design and implement analytical databases using Azure SQL Database, Azure Synapse Analytics, and an Azure Storage account. We used the data from our operational databases, supplemented with external data, log data, and streaming data, to populate these analytical databases to perform analytics on the data.

This section introduces Azure Data Factory, the tool to use when you need to enter data into analytical databases. The process of moving data from operational databases to analytical databases is called the **Extract, Transform, Load** (**ETL**) process. Azure Data Factory is the tool you use in Azure to implement this ETL process.

Both the ETL process and the Azure Data Factory tool deserve a book by themselves. This chapter is just a short introduction. It provides a link between the first two sections of this book. In this chapter, you will learn about the following topics.

- Data Factory overview
- Using the copy activity
- Using data flows

Technical requirements

You will need the following if you wish to successfully complete this chapter:

- A connection to the internet and a modern browser
- An Azure subscription with permission to create new services

Introducing Azure Data Factory

In *Chapter 8, Provisioning and Implementing an Azure Synapse SQL Pool*, you saw the modern data warehouse architecture as shown in *Figure 11.1*:

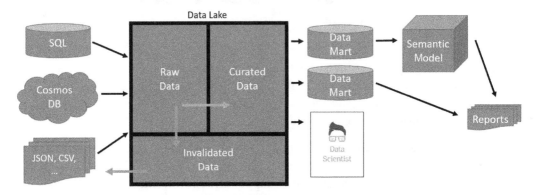

Figure 11.1 – Modern data warehouse architecture

All the arrows in *Figure 11.1* denote data movement activities. We need to export data from the operational databases and copy that data to the data lake's raw zone. We then process the newly added data and move the now transformed data into the curated zone of the data lake. From there, the data is transformed even more. Business rules are applied and the data is shaped into a star schema design to be stored in a data mart. We might then process an Analysis Services tabular model to serve the data to end users.

The entire process of moving data and transforming it is referred to as the ETL process. It sometimes involves copying data and, on other occasions, the data needs to be transformed. Depending on the situation, we can do multiple (independent) tasks at the same time or we can only start an activity when another has successfully finished because there is a dependency between the tasks at hand. For instance, you cannot validate data in the data lake when it has not yet been copied into the data lake.

We need all data movement and data transformations to be automated. We also need to be able to schedule all these actions. Azure Data Factory can do all that for us. It copies and transforms the data and allows us to create workflows that can be run based on various schedules.

Azure Data Factory comes in two versions: V1 and V2. V1 is a very limited version compared to the much newer V2. This chapter will, from here on, be about V2.

Introducing the main components of Azure Data Factory

Before getting our hands dirty with hands-on examples, let's get familiar with some terminology. We will do that by explaining the main components comprising Azure Data Factory. *Figure 11.2* shows the main components of Azure Data Factory and how they relate to one another:

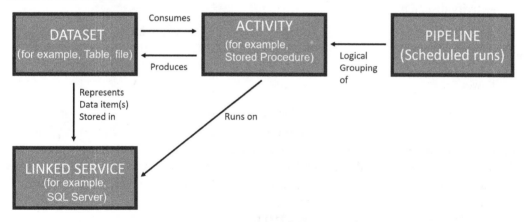

Figure 11.2 – Azure Data Factory components

The main components comprising Azure Data Factory are as follows:

- Activities
- Datasets

- Linked services
- Pipelines

Let's look at each of these components in some detail.

Understanding activities

An activity is a piece of work that needs to be completed. An activity is a single piece of work, a single step if you will, in a larger whole. You could also describe an activity as a task that needs to be performed. Basically, there are just two tasks: you either copy or transform data. Getting data into the cloud or, when it is cloud-born, getting it into the raw zone of a data lake or a staging database, often involves simply copying data. For a complete list of supported sources, have a look at `https://docs.microsoft.com/en-us/azure/data-factory/concepts-pipelines-activities#data-movement-activities`.

Once the data is ingested, it needs to be processed in a certain way. The data needs to be validated, cleansed, and transformed before it can be loaded into a dimensionally modeled data mart.

Data can be transformed using four different technologies. These technologies are as follows:

- Activity dispatch
- SSIS activity
- Data flow
- Power Query data wrangling

Activity dispatch means that Azure Data Factory lets another platform perform some transformation. You can write a SQL stored procedure in a Synapse Analytics SQL pool and let Data Factory execute this stored procedure. Or you can write Python or Scala code to transform the data and use Data Factory to run that code. Data Factory is merely used as a scheduler that executes tasks and activities at the right time. For the complete list of activities, refer to `https://docs.microsoft.com/en-us/azure/data-factory/concepts-pipelines-activities#data-transformation-activities`.

SQL Server Integration Services (**SSIS**) is the ETL tool that has been used a lot in on-premises ETL for the last 15 years. A lot of data transformations have been developed using SSIS. To leverage the investments made in SSIS in the past when moving to the cloud, you can run your SSIS packages using Data Factory.

Data flow is the part where Data Factory is doing the work by itself. It is a graphical environment, much like the old SSIS, where you can develop data transformations. It is a no-code development tool where you configure predefined steps to achieve the required data transformations.

The fourth option, Power Query data wrangling, leverages the Power Query mashup engine. This is the same engine as Power Query from Power BI and the Power BI data flows.

You do not have to choose just one of the options. An entire ETL process comprises a lot of steps. For each step, you can choose one of the activity types mentioned. For instance, you can use Scala on Databricks to validate incoming data in the data lake. You might then use a combination of Python and SQL that you run in a Synapse Analytics workspace. Choosing one technology to do all your ETL might be better from a maintenance perspective. However, different steps have different requirements and choosing the best tool for the job lets you take full advantage of the cloud and might make your solution more cost-efficient.

Understanding datasets

Activities are about what you do with your data and where you do it. Datasets define the data that you actually perform the activities on. Activities have input. That is the data the activity is supposed to work on. That data is stored somewhere; it has a location. The data is also stored in a certain format that your activity needs to know about. We learned about different file formats, such as AVRO and ORC. When you work with files, you need to know whether they are compressed, and if so, how.

Next to the input, an activity will write the transformed data into a new location. An activity has an output. Both inputs and outputs of activities are defined as datasets.

This may sound formal and abstract, but it basically boils down to the fact that a dataset is a table in a relational database or a file or folder in the data lake. It can also be a Cosmos DB container. The exact form the dataset takes depends on the data store the data resides in.

Basically, a dataset is like a view that represents data. It is the data you want to work with, or it is the result of an activity. In *Figure 11.2*, we see this in the arrows drawn between the dataset and the activity. An activity has an input; it consumes data. The activity then produces a dataset, a copy of the data, or a transformed version of the data. This is the output of the activity.

Understanding linked services

A linked service is just a connection string. A connection string is all the information you need to connect to a data platform. When you work with a SQL database, the connection string holds the name of the SQL server, the database name, and the security information needed to connect to that database. For a file, the connection string might be just the filename with the entire path.

Azure Data Factory needs linked services for both activities and datasets. To start with the latter, a dataset might be a table in a database. You need a linked service to connect to that database. The dataset is the logical representation of the data. The accompanying linked service defines the location where the table is stored and the security information to connect to it. This is no different for a file. A dataset can be a file in your data lake. The linked service holds the information of your Azure storage account and the security settings you want to use to connect to it. In other words, a linked service is what was called a connection manager in SSIS.

This use of a linked service is drawn in *Figure 11.2* as an arrow between a dataset and a linked service.

When you use activity dispatch, a data platform other than Azure Data Factory itself will execute the activity. A SQL stored procedure needs to be run on a SQL server. A linked service defines the SQL server on which to run the activity. Again, the linked service is just the connection information defining the SQL server on which to run the activity and the security settings to run it with. You can see an arrow pointing from the activity to the linked service in *Figure 11.2* to represent this use of linked services.

Understanding pipelines

A pipeline is a series of activities linked together. An activity was explained as a single step in a larger whole. A pipeline links the steps in a certain way to create that larger whole. You may want to run multiple activities in parallel, taking advantage of the hardware you have. Or you may need to run steps serially because there are dependencies between activities, such as loading dimension tables before loading fact tables.

Pipelines can also be regarded as units of work. You execute a pipeline, not an individual activity. A pipeline can consist of a single activity or multiple activities. You always execute (run) the pipeline as a whole. This can be done manually or based on a trigger. Various kinds of triggers exist. One example is a wall-clock trigger that allows you to schedule a pipeline to run at a specified time of the day.

Another word that comes to mind when describing a pipeline is workflow. A workflow is a series of steps that need to be run in a certain order, like a pipeline being a series of activities executed in a certain order.

Activities, datasets, linked services, and pipelines are the main components to be aware of. However, it makes sense to also introduce triggers and integration runtimes.

Understanding triggers and integration runtimes

Azure Data Factory is an automation tool. You define what exactly ETL means to you by creating pipelines. These pipelines need to be executed. You can manually execute a pipeline. You can use the .NET SDK, Azure PowerShell, the REST API, or the Python SDK to trigger a pipeline run yourself.

Data Factory also provides three types of triggers that automatically execute a pipeline. These triggers are as follows:

- Schedule trigger
- Tumbling window trigger
- Event-based trigger

A schedule trigger invokes a pipeline on a wall-clock schedule. The pipeline will always run at the same time. These types of triggers are often used in more "classic" BI architectures, where we extract data from operational systems, such as a CRM database, every night. A tumbling window trigger operates on a periodic interval, while also retaining a state. Tumbling window triggers process your data in time slices, which may be far more efficient with streaming data and IoT scenarios. Another scenario might be where you extract new data from a database based on timestamps and upload that new data to your data lake where you store it in time-sliced folders. Event-based triggers, as the name implies, react to events. You can think of an event as a file arriving in a storage account.

As soon as a trigger starts executing a pipeline, the tasks contained in the pipeline will need to be run. The hardware used to run the activities is defined by an integration runtime. There are three types of integration runtimes: Azure, self-hosted, and Azure-SSIS.

The Azure integration runtime uses hardware managed by Microsoft. You use this when all your data is in Azure cloud services such as Cosmos DB, SQL Database, and Azure Storage. A self-hosted integration runtime can be installed on your own servers. These can be Azure VMs or on-premises servers. This allows you to create hybrid scenarios with on-premises data sources. It also allows you to work with data sources that you need to install special drivers for. When you want to run SSIS packages, you need to choose the Azure SSIS integration runtime.

Figure 11.3 brings all these components together in an overview:

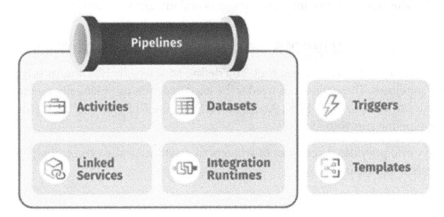

Figure 11.3 – Azure Data Factory overview

To recap what we have learned so far, a pipeline is a workflow that can be executed manually or by a trigger. The workflow consists of one or more activities. Activities that copy or transform data have an input and an output that are defined by datasets. Data Factory connects to the data source or service using linked services. An activity uses an integration runtime, which determines the infrastructure used.

The one new component in *Figure 11.3* is the template. Azure Data Factory provides templates that you can use to start building pipelines. You can add your own templates to the predefined templates. This allows you to reuse a starting point for easier and quicker development.

Now that you have learned about the main components of Azure Data Factory, let's look at how it works by means of two simple examples. We will first use the copy activity to ingest data into our data lake. We will then use a data flow to transform some data to finally use activity dispatch to load data into Azure Synapse Analytics.

Using the copy activity

You can create pipelines and activities from two different places. The first is Azure Data Factory itself. The second place is Azure Synapse Analytics. The "integrate" part of Synapse Analytics is actually the same as Azure Data Factory. ETL is such an integral part of analytical databases that Microsoft brings everything involved in developing analytical databases into a single environment. We have the serverless SQL pool and the Spark pool to perform advanced transformations on data when necessary. And we have pipelines with all the different sorts of activities to create the entire workflow. Because this book is about databases and not ETL by itself, we will work from Synapse Analytics.

We will first copy a single table. We will then move on to copy all tables.

Copying a single table to the data lake

We will begin by copying the customer table from the `Northwind` database into the data lake created in *Chapter 10, Designing and Implementing a Data Lake Using Azure Storage*:

1. Navigate to the **Home** blade of the Synapse workspace.

2. Click on **Open** inside the **Open Synapse Studio** tile to open Synapse Studio.

   There are multiple approaches you can take here. A logical first step would be to define the datasets and linked services that you plan to work with.

3. In the **Synapse Studio** menu, click on **Data** (see *Figure 11.4*).

4. Click on the plus icon (**Add new resource**) beside **Data**.

5. Select the **Integration dataset** option:

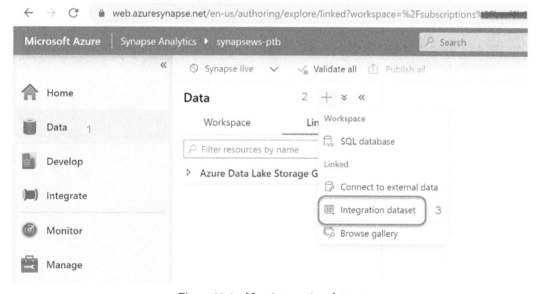

Figure 11.4 – New integration dataset

6. Select **Azure SQL Database** and then click **Continue**.

7. Insert `NW_Customers` as the name.

8. In the drop-down box under **Linked service,** click on the **New** option.

9. Use the following settings for the **New linked service** dialog (Azure SQL Database) and leave the other settings as their defaults:

Name: Northwind.

Azure subscription: The subscription you used to create a database in *Chapter 4, Provisioning and Implementing an Azure SQL DB.*

Server name: dbdesignbook

Database name: Northwind

User name: Your name (or the user you created in *Chapter 4, Provisioning and Implementing an Azure SQL DB*, when provisioning the SQL server).

Password: The password you created in *Chapter 4, Provisioning and Implementing an Azure SQL DB.*

Click on **Test connection** to test that all the settings are correct.

Click on **Create**.

10. Back in the **Set properties** pane, select **dbo.Customer** in the drop-down list under **Table name** and then click **OK**.

Your screen should now look similar to *Figure 11.5*:

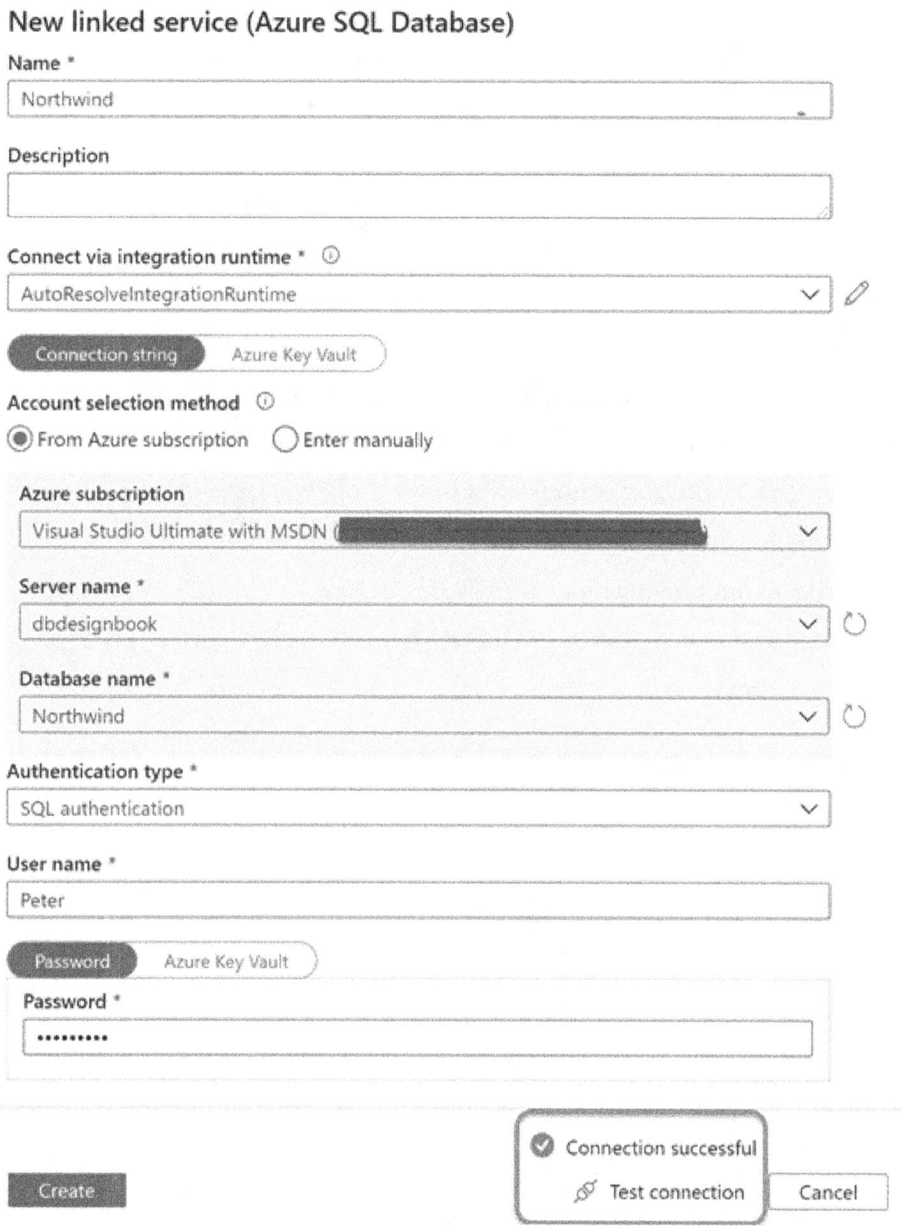

New linked service (Azure SQL Database)

Name *

> Northwind

Description

Connect via integration runtime * ⓘ

> AutoResolveIntegrationRuntime   ∨   ✎

(**Connection string**   Azure Key Vault)

Account selection method ⓘ

◉ From Azure subscription   ◯ Enter manually

Azure subscription

> Visual Studio Ultimate with MSDN (   )   ∨

Server name *

> dbdesignbook   ∨   ↻

Database name *

> Northwind   ∨   ↻

Authentication type *

> SQL authentication   ∨

User name *

> Peter

(**Password**   Azure Key Vault)

Password *

> •••••••••

[Create]   ✅ Connection successful

                        ✑ Test connection   [Cancel]

Figure 11.5 – New linked service

You just created a linked service to the Northwind database on the server called
dbdesignbook. You named this linked service Northwind to make clear that it is
merely a reference to the Northwind database. You also created a dataset called NW_
Customers that references the dbo.Customer table in the Northwind database.

With the source-linked service and dataset created, you also need a linked service and dataset for the copy of the dbo.Customer table that you need to write to the data lake:

1. Create a new integration dataset by clicking on the plus icon beside **Data** (see *Figure 11.4*).

2. Choose **Azure Data Lake Storage Gen2** and then click **Continue**.

3. Choose **DelimitedText** and then click **Continue**.

4. Enter CustomerCSV as the name.

5. Select **New** in the drop-down list under **Linked service**.

6. Use the following settings for the **New linked service** dialog (Azure Data Lake Storage Gen2) and leave the other settings as their defaults:

    a. **Name**: DesignBookDataLake.

    b. **Azure subscription**: The subscription you have been using so far.

    c. **Storage account name**: designbookdatalake.

    d. Click on **Test connection** to test that all your settings are correct. See *Figure 11.6*.

    e. Click on **Create**:

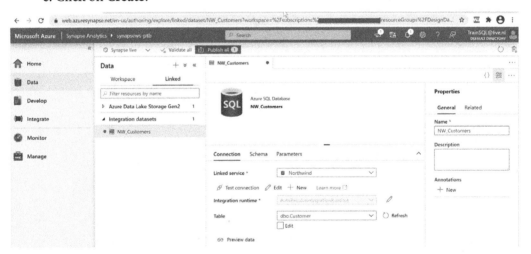

Figure 11.6 – New data lake linked service

7. Back in the **Set properties** pane, select northwind/raw under **File path** as in *Figure 11.7* and then click **OK**:

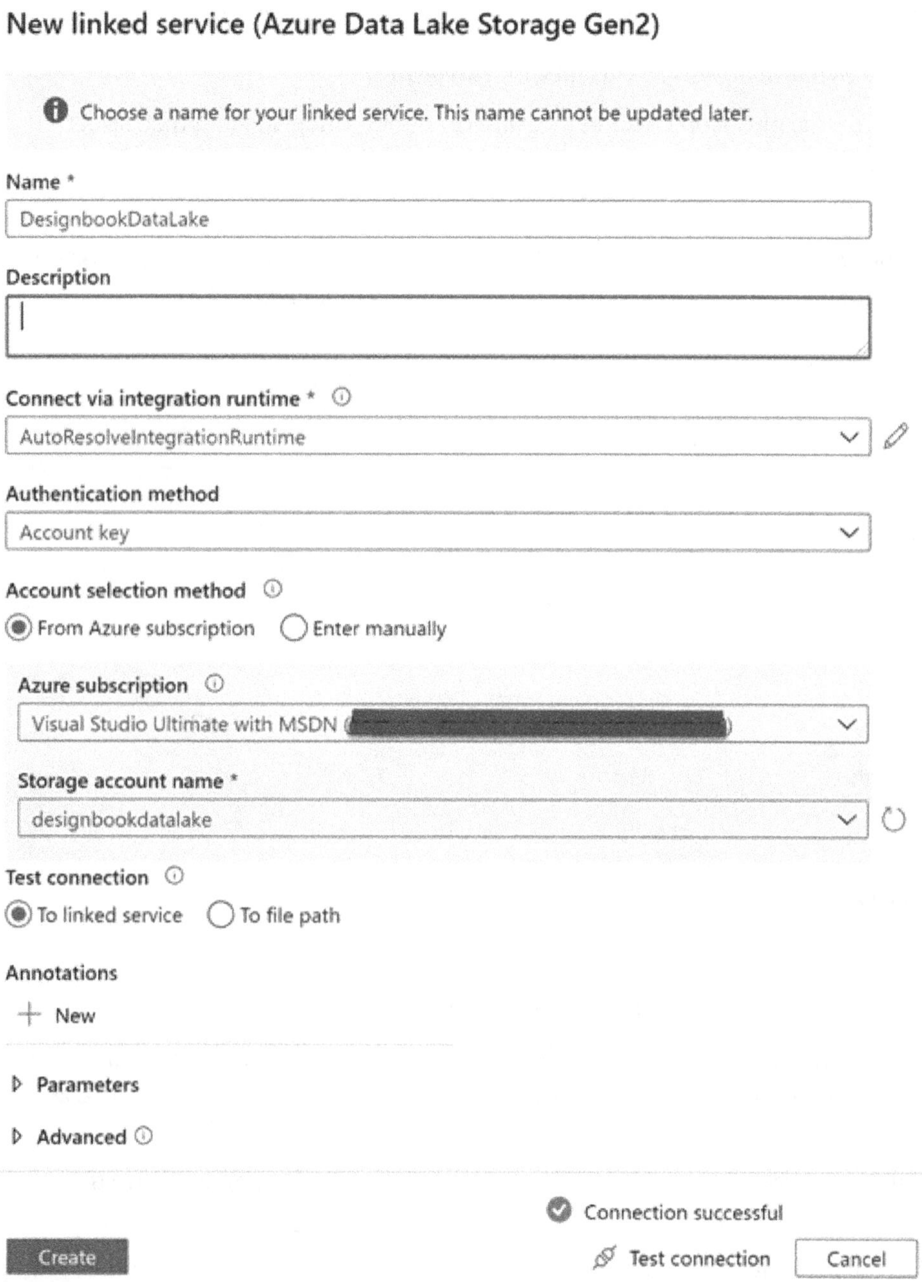

New linked service (Azure Data Lake Storage Gen2)

ⓘ Choose a name for your linked service. This name cannot be updated later.

Name *

```
DesignbookDataLake
```

Description

```
|
```

Connect via integration runtime * ⓘ

```
AutoResolveIntegrationRuntime                                    ⌄  ✎
```

Authentication method

```
Account key                                                      ⌄
```

Account selection method ⓘ

◉ From Azure subscription   ◯ Enter manually

Azure subscription ⓘ

```
Visual Studio Ultimate with MSDN (████████████████████████)      ⌄
```

Storage account name *

```
designbookdatalake                                               ⌄  ↻
```

Test connection ⓘ

◉ To linked service   ◯ To file path

Annotations

+ New

▷ Parameters

▷ Advanced ⓘ

✓ Connection successful

🖉 Test connection   Cancel

Create

Figure 11.7 – New data lake dataset

Notice that we choose to store the export from the Northwind database as CSV files in our data lake.

With both a source and a sink set up, it is time to actually copy the data:

1. In the **Synapse Studio** menu, click on **Integrate**.

2. Click on the plus icon (**Add new resource**) beside **Integrate** and then select **Pipeline**.

3. Name the pipeline CopyCustomers.

4. Under **Activities**, open the **Move & transform** section and then drag a **Copy data** transform to the canvas just right of the activities (see *Figure 11.8*):

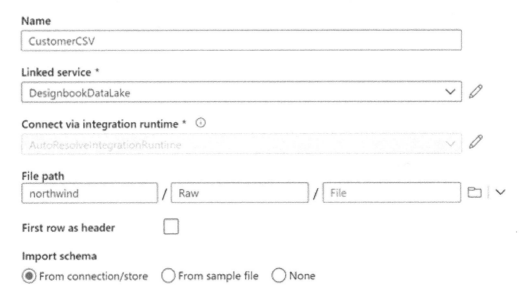

Figure 11.8 – New pipeline with a copy activity

5. Make sure that copy activity is selected and that you see several tabs with settings at the bottom of your screen. Under **General**, enter CopyCustomerData as the name.

6. On the **Source** tab, select **NW_Customers** for **Source dataset** and leave the other settings as they are.

7. On the **Sink** tab, select **CustomerCSV** for **Sink dataset** and leave the other settings as they are.

It is beyond the scope of this book to discuss all the settings and options you can choose from here. Suffice to say that this is all you need to copy data from the SQL server into your data lake. All you need is to execute the pipeline. In other words, you need a trigger. But before you can execute the pipeline, you need to save it.

When setting up pipelines and schedules, it is a good idea to set up a code repository where you can save your work. From here, you can also collaborate with co-workers. Additionally, you can set up versioning. And another thing: you can save your work into your code repository whenever you feel like it.

Without setting up a repository, you need to publish your work to the Synapse workspace before you can execute it. You should see a blue button, **Publish all**, at the top of your screen (see *Figure 11.9*). You can only publish work that is validated:

8. Click on **Publish all** at the top of the screen and then click on **Publish** in the dialog that opens on the right-hand side of your screen.

9. With the pipeline selected, click on **Add trigger** (see *Figure 11.9*) and select **Trigger now** to execute the pipeline. Click on **OK** in the dialog that opens on the right-hand side of your screen:

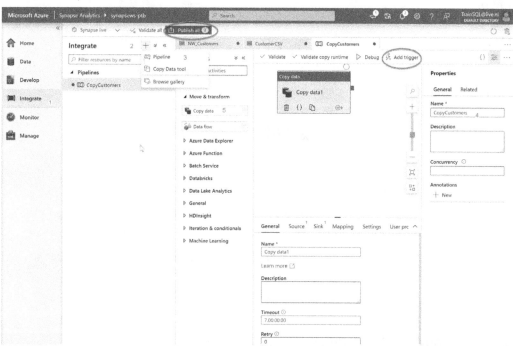

Figure 11.9 – New pipeline with a copy activity

10. In the **Synapse Studio** menu on the left-hand side of your screen, click on **Monitor**.

11. Select **Pipeline runs**. Your screen should look as in *Figure 11.10*:

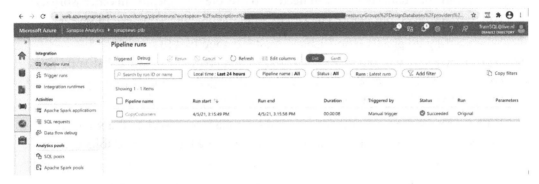

Figure 11.10 – Monitoring pipeline runs

12. In the Azure portal, navigate to the **Home** blade of the data lake storage account, `designbookdatalake`.

13. Open the Storage Explorer from the left-hand side menu.

14. Verify that your pipeline added a file with the name `dbo.Customer.txt`.

Even though copying a table from the database into the data lake is easy enough like this, you do not want to repeat all these steps for all other tables that we still need to extract from the `Northwind` database. You want a pipeline that copies all (relevant) tables into the data lake. So let's make such a pipeline.

Copying all tables to the data lake

Let's create a second pipeline that copies all tables to the data lake:

1. Navigate back to Synapse Studio.

2. Click on **Integrate** in the left-hand side menu of Synapse Studio and create a new pipeline. Name it `CopyAllTables`.

3. Open the **General** section under **Activities** and drag a **Lookup** activity to the design canvas.

4. On the **Settings** tab of the **Lookup** activity, click on **New** beside the drop-down list for **Source Dataset**.

5. Select **Azure SQL Database** and then click **Continue**.

6. Name the new dataset, `AllTables`, and then select **Northwind** as the linked service. Leave the **Table name** setting empty and then leave the **Import schema** setting as **None**. Click on **OK**.

7. Change the **Use query** setting in **Query** and provide the following SQL statement:

```
SELECT
                '[dbo].' + QUOTENAME(name) AS TableName
FROM sys.tables
WHERE type = 'U';
```

This statement gets a list of all user tables from the `Northwind` database, assuming all tables are within the **dbo** schema.

8. Make sure that the **First row only** setting is *not* selected.

9. Rename the **Lookup** activity to `GetAllTables` on the **General** tab of the activity.

The **Lookup** activity retrieves a list of table names based on your own logic. A **ForEach** activity can perform activities on each item coming from the lookup.

10. Open the **Iteration & conditionals** section under **Activities** and drag a **ForEach** activity to the canvas.

11. Connect both activities by dragging the green dot on the right-hand side of **Lookup** over to the **ForEach** activity.

12. Rename the **ForEach** activity to `CopyEachTable` on the **General** tab of the activity.

13. Go to the **Settings** tab of the **ForEach** activity and click on the **Items** textbox. This should make a link visible that says **Add dynamic content [Alt+P]**. Click on this link.

14. Add the following code to the dynamic content and then click **Finish** (see *Figure 11.11*):

```
@activity('GetAllTables').output.value
```

Here is the output:

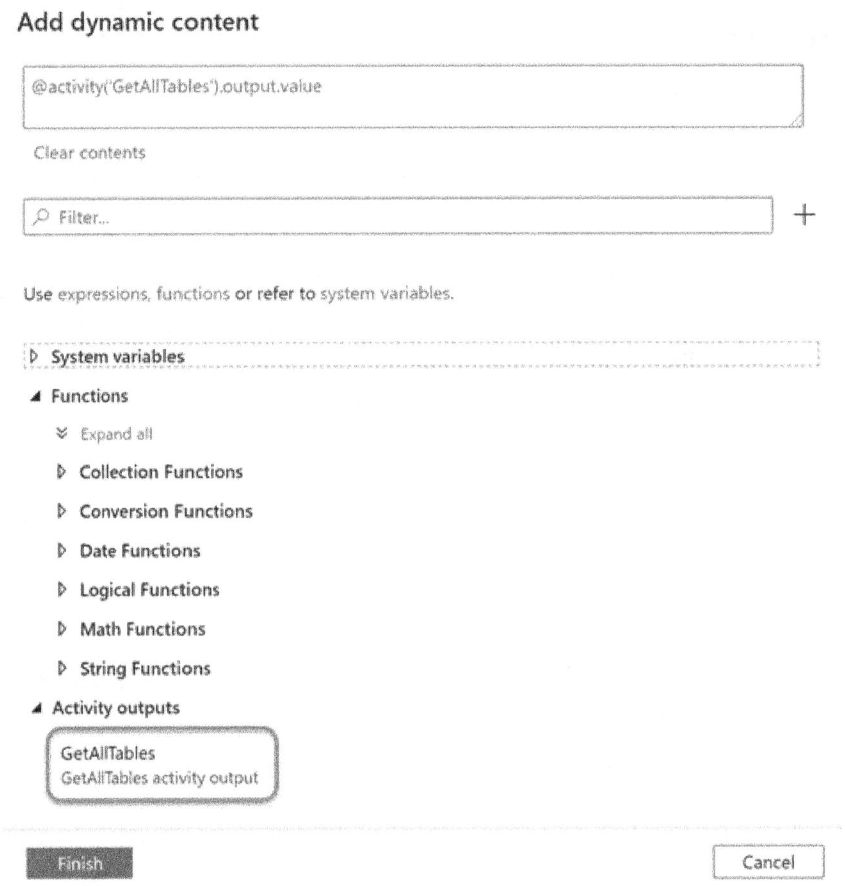

Add dynamic content

@activity('GetAllTables').output.value

Clear contents

🔍 Filter...   +

Use expressions, functions or refer to system variables.

▷ **System variables**

◢ **Functions**

  ⩛ Expand all

  ▷ **Collection Functions**

  ▷ **Conversion Functions**

  ▷ **Date Functions**

  ▷ **Logical Functions**

  ▷ **Math Functions**

  ▷ **String Functions**

◢ **Activity outputs**

  GetAllTables
  GetAllTables activity output

Finish   Cancel

Figure 11.11 – Adding dynamic code

This code makes sure that the **ForEach** activity gets passed through all the table names from the **Lookup** activity. It will iterate through all the names (all the items) and perform some activities on them. The activities it will execute are defined in the **Activities** tab of the settings of the **ForEach** activity.

15. Select the **Activities** tab of the settings of the **ForEach** activity.

16. Click on the pencil icon beside the **No activities** test. This will open the design canvas for the activities that need to be performed for each table.

17. Add a copy activity to the newly opened design canvas.

18. On the **Source** tab of the settings for the copy activity, select **AllTables** to be the source dataset.

19. Select the **Query** option beside the **Use Query** setting and use the following query:

```
SELECT * FROM @{item().TableName}
```

This query uses a table name from the **ForEach** activity by referring to the item collection of the **ForEach** activity. This dynamically builds a SQL SELECT statement that reads all the data from a table. It gets the table name via the item collection that is filled by the query in the **Lookup** activity.

We now need to write the tables to our data lake.

20. On the **Sink** tab, click **New** beside **Sink dataset** to create a new dataset.

21. Select **Azure Data Lake Storage Gen2** and then click **Continue**.

22. Select **DelimitedText** and then click **Continue**.

23. Select **DesignbookDataLake** as the linked service.

24. Select the **Raw** folder from the Northwind container as the file path.

25. Back on the **Sink** tab of the **Copy** activity settings, open the just-created dataset.

26. Change the dataset to be as in *Figure 11.12*, where you add northwind/Raw as the first type of the file path and then add dynamic content to the second part to create a new folder for each table you copy. The dynamic content is @{item(). TableName}:

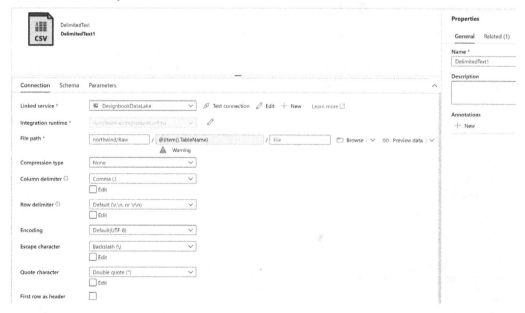

Figure 11.12 – Sink configuration

That should be it. You have read all the tables from the database. You then have a **ForEach** activity that executes a copy activity for each table so that data will be copied into a folder with the same name as the table. All you need to do is perform a test:

27. Click **Publish all** and then click **Publish** again.

28. Click **Add trigger** followed by **Trigger now**. Click on **OK** in the dialog that appears on the right-hand side of your screen.

29. In the **Synapse Studio** menu on the left-hand side of your screen, click on **Monitor**.

30. Select **Pipeline runs**. Check that your pipeline ran successfully.

31. In the Azure portal, navigate to the **Home** blade of the data lake storage account, `designbookdatalake`.

32. Open the Storage Explorer from the left-hand side menu.

33. Verify that your pipeline created a folder for each table and added a file to the folder.

You just successfully exported an entire database to your data lake. The next step is to work with that data. In the next section, we will implement a simple data flow to manipulate the data.

Implementing a data flow

Now that we have data in our data lake, let's transform the data and load it into our Synapse SQL pool that we created in *Chapter 8, Provisioning and Implementing an Azure Synapse SQL Pool*. As an example, we will create a pipeline that uses data flows to do a full load of the `dimProduct` table. You created the `dimProduct` table using the following SQL statement:

```
CREATE TABLE dbo.dimProduct
(
  ProductKey              int            NOT NULL
, SourceKey               int            NOT NULL
, ProductName             nvarchar(40)   NOT NULL
, CategoryName            nvarchar(15)   NOT NULL
, QuantityPerUnit         nvarchar(20)   NULL
, UnitPrice               money          NULL
, ReorderLevel            smallint       NULL
, Discontinued            bit            NOT NULL
, Supplier_CompanyName    nvarchar(40)   NOT NULL
```

```
, Supplier_ContactName          nvarchar(30)        NULL
, Supplier_ContactTitle         nvarchar(30)        NULL
, Supplier_Address              nvarchar(60)        NULL
, Supplier_City                 nvarchar(15         NULL
, Supplier_Region               nvarchar(15)        NULL
, Supplier_PostalCode           nvarchar(10)        NULL
, Supplier_Country              nvarchar(15)        NULL
, Supplier_Phone                nvarchar(24)        NULL
)
WITH
(
            DISTRIBUTION = ROUND_ROBIN
          , CLUSTERED INDEX (ProductKey ASC)
);
```

This table has columns from the Product table in Northwind, but also from the Supplier and ProductCategory tables. We need to combine the data from these three tables in the data flow we are going to create. We also need to generate values for the surrogate key, ProductKey.

Let's start by reading product data from the data lake using a new data flow:

1. If necessary, open Synapse Studio.
2. In the **Synapse Studio** menu on the left-hand side of your screen, click on **Develop**.
3. Click on the plus icon and then click on **Data flow** in the popup that appears.
4. Enable data flow debugging on top of the designer. This starts a cluster that allows you to see data previews, as well as reading metadata.
5. Close any popups that appear and click on the area that shows the text **Add source**.
6. On the **Source settings** tab, enter ProductTable as the output stream name.
7. Click on the **New** button beside **Dataset** to create a new dataset that will refer to the product CSV file in the data lake.
8. Select **Azure Data Lake Storage Gen2** and then click **Continue**.
9. Click on **Delimited text** and then click **Continue**.
10. Enter Product_in_DataLake for the name of the dataset.
11. Select the **DesignbookDataLake** linked service that you created earlier that references your data lake.

12. Select the [dbo].[Product] folder in the raw zone of the data lake under the **File path** setting as in *Figure 11.13*, and then click **OK**.

13. Select the **Projection** tab and then click on **Import projection**. This reads the metadata from the files in the selected folder:

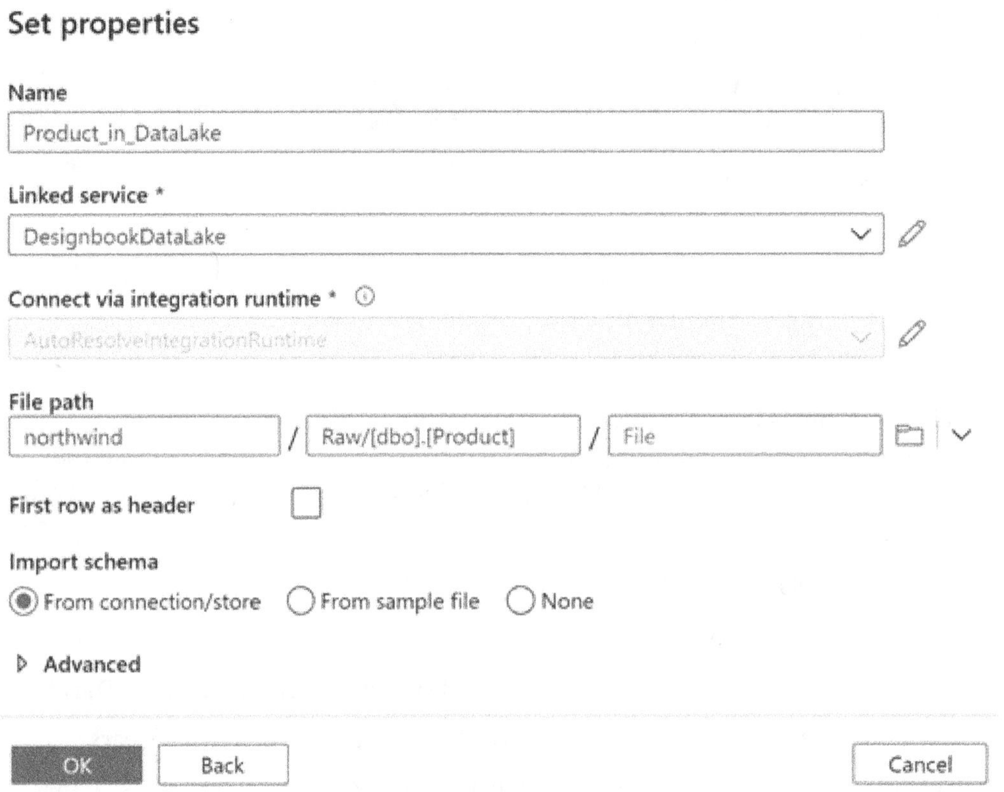

Set properties

Name

| Product_in_DataLake |

Linked service *

| DesignbookDataLake   ∨ | 🖉 |

Connect via integration runtime * ⓘ

| AutoResolveIntegrationRuntime   ∨ | 🖉 |

File path

| northwind | / | Raw/[dbo].[Product] | / | File | 📁 | ∨ |

First row as header   ☐

Import schema

◉ From connection/store   ◯ From sample file   ◯ None

▷ Advanced

| OK | | Back | | Cancel |

Figure 11.13 – Product_in_DataLake dataset

The settings so far make sure that you read all data in the [dbo].[Product] folder. There are a couple of things you need to do before you can load this data into the SQL pool called NW_Sales_DM. The first thing to do is to generate surrogate key values. The data flow has the **Surrogate key** schema modifier to do just that:

1. Click on the little plus icon in the bottom right-hand corner of the **ProductTable** source you just created.

2. Select **Surrogate key** from the schema modifiers to add a new task to the data flow.

3. On the **Surrogate key settings** tab, enter `AddedProductKey` for the **Output stream name** setting.

4. Enter `ProductKey` for the **Key column** setting.

The **Surrogate key** schema modifier numbers the rows starting with 1. When you do incremental loads, you might want to write a query to the `dimProduct` table to retrieve the largest `ProductKey` value currently in the table. You then add a **Derived column** schema modifier where you add the retrieved maximum value to the value generated by the **Surrogate key** schema modifier. With a full load, you can start the numbering at 1, as this is the default.

Next, you need to add category information to the data flow:

1. Click on **Add source** below the **ProductTable** source to add another data source to the data flow.

2. Set the output stream name of the newly added data source to `ProductCategoryTable`.

3. Create a new dataset similar to the one you created in *steps 7* to *12*, selecting the **ProductCategory** folder instead of the **Product** folder in the data lake to create a source that reads in category data.

4. Click on **Import projection** on the **Projection** tab.

    After adding a second data source, you need to join the two data sources.

5. Click on the little plus icon in the bottom right-hand corner of the **AddedProductKey** step that you created in *steps 14* and *15*.

6. Select **Join** in the popup that appears.

7. Enter `JoinedCategories` as the output stream name.

8. Select **ProductCategoryTable** in the drop-down list for the **Right stream** setting.

9.   Configure an inner join on _col0_ for both tables, as in *Figure 11.14*:

Figure 11.14 – Data flow join

Now that you have joined two tables, you need to add columns from a third table, Supplier, to your data.

10. Add another data source to the data flow that reads data from the [dbo] . [Supplier] folder in the data lake similar to how you added the Product and ProductCategory data. Name this data source SupplierTable.

11. Add a second join operator after the **JoinedCategories** step and call it JoinedSuppliers. Use the newly added **SupplierTable** table for the **Right stream** setting. Join on **ProductTable@_col2_ and _col0_.**

Now that you have joined all the necessary tables, it is time to realize that you don't use all the columns in the dimProduct table. So, you need to select the used columns and we might as well add readable metadata to the columns while doing so:

1.   Click on the little plus icon in the bottom right-hand corner of the **JoinedSuppliers** step you just created.

2.   Choose **Select** from the schema modifiers to add a new task to the data flow.

3. Make a mapping as in the table shown here. Remove all the columns not mentioned in the table:

JoinedSuppliers column	Name as
ProductTable@_col0_	SourceKey
ProductTable@_col1_	ProductName
ProductTable@_col4_	QuantityPerUnit
ProductTable@_col5_	UnitPrice
ProductTable@_col8_	ReorderLevel
ProductTable@_col9_	Discontinued
ProductKey	ProductKey
ProductCategoryTable@_col1_	CategoryName
SupplierTable@_col1_	Supplier_CompanyName
SupplierTable@_col2_	Supplier_ContactName
SupplierTable@_col3_	Supplier_ContactTitle
SupplierTable@_col4_	Supplier_Address
SupplierTable@_col5_	Supplier_City
SupplierTable@_col6_	Supplier_Region
SupplierTable@_col7_	Supplier_PostalCode
SupplierTable@_col8_	Supplier_Country
SupplierTable@_col9_	Supplier_Phone

Table 11.1 – Column mapping

With the proper columns selected, it is time to write the result to our table in the SQL pool, NW_Sales_DM:

1. Click on the little plus icon in the bottom right-hand corner of the **Select** step you just created.

2. Select **Sink** from the list that appears to add a new task to the data flow.

3. Name the new sink dimProduct.

4. On the **Sink** tab, click on **New** beside the **Dataset** setting.

5. Select **Azure Synapse Analytics** and then click **Continue**.

6. Name the dataset SynapseTableDimProduct.

7. Click on **New** to create a new linked service to the SQL pool. Use the following settings:

   a. Select your subscription.

   b. Select your Synapse workspace under the name of the server.

   c. Select NW_Sales_DM under the name of the database.

   d. Enter your name and the password you created under **Username** and **Password**.

8. Test the connection and click on **Create** once everything is set correctly.

9. Select **dbo.dimProduct** under **Table name** and then click on **OK**.

10. Click on the **Settings** tab and then select **Truncate** for the **Table action** setting.

11. You can check the column mappings on the **Mappings** tab.

The data flow is ready. All you need to do is create a pipeline to run the data flow:

12. In the **Synapse Studio** menu on the left-hand side of your screen, click on **Integrate** and then create a new pipeline. Give the pipeline an appropriate name.

13. Under **Activities**, open **Move & transform** and then drag a **Data flow** activity to the designer.

14. On the **Settings** tab, under **Data flow**, select the data flow you just created.

15. Select your Synapse workspace default storage for the **Staging linked service** setting.

16. Select the **Raw** folder as the staging storage folder.

A data flow copies data that is inserted into Synapse to data lake storage first so that it can use PolyBase to insert the data into the Synapse table. As you learned in *Chapter 8, Provisioning and Implementing an Azure Synapse SQL Pool*, PolyBase is the most efficient way to import data into Synapse. The last two settings that you provide configure the storage that the data flow can use for this purpose.

Everything should now be set. You are ready to load data into Synapse:

1. Click on **Publish all**, and then click on **Publish** in the dialog that opens.

2. Click on **Add trigger** and then on **Trigger now** in the popup that appears.

3. Navigate to the **Monitor** page of Synapse Studio to verify that your pipeline ran successfully. This may take some time.

4. Navigate to the **Data** page of Synapse Studio and verify that you have data in the **dimProduct** table in the **NW_Sales_DM** database. See *Figure 11.15*:

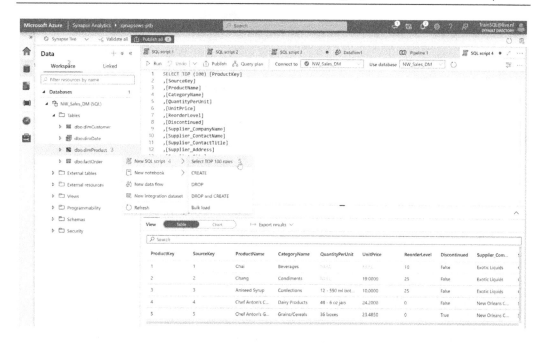

Figure 11.15 – dimProduct

Data flow is a no-code ETL environment for moving data through your modern data warehouse environment. You just used it to load a data warehouse dimension table.

You can also write code to do ETL. The next section shows how to load data to another table using SQL code.

Executing SQL code from Data Factory

Azure Data Factory, and through Data Factory, Synapse integration, lets you choose the most suitable way to load data into a Synapse SQL pool. You just used the code-less option by implementing a Data Factory data flow. You already imported data into your Synapse SQL pool using T-SQL code in *Chapter 8, Provisioning and Implementing an Azure Synapse SQL Pool*. Let's have a quick look at how to automate the option where you use T-SQL code using Synapse pipelines.

In the following section, we will assume that you have implemented the `dimDate`, `dimCustomer`, `dimProduct`, and `factOrder` tables, as discussed and created in *Chapter 8, Provisioning and Implementing an Azure Synapse SQL Pool*. We will create a stored procedure in the SQL database and execute it using a pipeline activity. The stored procedure will do a full load of the `factOrder` table based on the exports of the `dbo.Order` and `dbo.OrderDetail` tables you created in the section regarding the copy activity:

1. If necessary, open Synapse Studio. Also, if necessary, resume the Synapse SQL pool called `NW_Sales_DM`.

2. Navigate to the **Data** tab of Synapse Studio.

3. Hover with the mouse over the **NW_Sales_DM** database, click on the three dots, and then create a new empty SQL script.

4. Open the SQL script, **setup.sql**, from the downloads and copy the contents into the empty SQL script.

5. *Step 1* of the script creates a master key. You can skip this step if you are working with the same SQL pool you created in *Chapter 8, Provisioning and Implementing an Azure Synapse SQL Pool*. Otherwise, select the code under *step 1* and click on **Run**.

6. *Step 2* of the script creates a database scoped credential to provide the Synapse SQL pool with access to your data lake. We created an Active Directory application in *Chapter 8, Provisioning and Implementing an Azure Synapse SQL Pool*, to get access. In this example, we use the access key of the storage account. Replace the < PASTE YOUR SECRET KEY HERE > text used in the script with the access key of your data lake storage account. Select the code under *step 2* and then click on **Run**. (You can find the key for your storage account by navigating to the storage account using the Azure portal. You will find it under the **Access keys** menu option.)

7. *Step 3* of the script creates an external data source and an external file format. Make sure to change the URL in the external data source to match your container name and storage account name. Use the container that you exported all the data from `Northwind` to in the section regarding the copy activity. Select the code under *step 3* and then click on **Run**. (If you get an error saying **Type with name 'AzureStorage' already exists**, you can append a 1 to the name. Make sure you do the same in subsequent steps.)

8. *Step 4* of the script creates two external tables. The `factOrder` table that we are going to load has columns originating in either the `Order` table or the `OrderDetail` table. The external tables reference the folder the data is in. Select the code under *step 4* and then click on **Run**.

9. To check whether the tables have been created correctly, open the **External tables** folder of the database on the left-hand side of your screen. Hover with your mouse over one of the tables you just created and click on the action button that appears. In the popup, click on **New SQL script** and then on **Select top 100 rows**. You should see your data.

10. The final step creates a stored procedure that uses a CTAS statement to load data from the external data into a temporary table in Synapse. It then uses an `INSERT SELECT` statement that fetches keys from pre-loaded dimension tables and inserts the data into the `factOrder` table. Notice that we didn't load all the dimension tables in the examples until now. The stored procedure translates missing keys into `-1` and leaves the tables we haven't loaded commented out. Select the last part of the script and execute it to create the stored procedure.

With the setup done, you are ready to use a pipeline with a stored procedure activity to load data into the `factOrder` table:

1. Navigate to the **Integrate** page of Synapse Studio.

2. Click on the plus icon and then on **Pipeline** to create a new pipeline.

3. Open the **General** section under **Activities** and drag a **Stored procedure** activity to the pipeline canvas.

4. With the **Stored procedure** activity selected, click on the **Settings** tab at the bottom of the screen.

5. Select the **SynapseAnalytics** linked service you created earlier for the linked service setting.

6.   Select the stored procedure you created in *step 10* for the stored procedure. Refer to *Figure 11.16*:

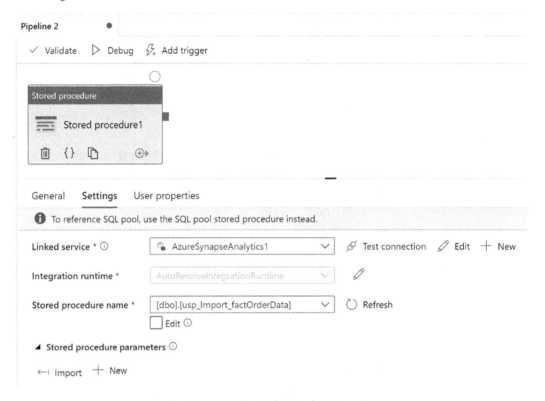

Figure 11.16 – Stored procedure activity

All that is left to do is to run the pipeline:

1.   Click on **Publish all** and then click on **Publish** in the dialog that opens.

2.   Click on **Add trigger** and then on **Trigger now** in the popup that appears.

3.   Navigate to the **Monitor** page of Synapse Studio to verify that your pipeline ran successfully.

4.   Navigate to the **Data** page of Synapse Studio and verify that you have data in the factOrder table in the NW_Sales_DM database.

You just successfully loaded data from the data lake into your Synapse SQL pool fact table. Note that in this chapter, we didn't really use a data lake. We merely used our data lake storage as a staging area for data that got exported from the Northwind database to be loaded into a Synapse data mart.

Summary

In this chapter, you used the Azure Data Factory functionality from the Synapse Studio environment. You used the copy activity to export data from an operational database and store it in the raw section of the data lake. You then used both a data flow and T-SQL to pick up data from the data lake to load it into the data mart. The data is now ready for analytics. Using, for instance, Power BI, we can start creating reports and dashboards using the prepared data.

That concludes this book. You learned about different data stores and how to design those data stores for optimal performance while minimizing the cost. I hope this book will prove advantageous when it comes to using the knowledge acquired in real-world scenarios.

Packt.com

Subscribe to our online digital library for full access to over 7,000 books and videos, as well as industry leading tools to help you plan your personal development and advance your career. For more information, please visit our website.

Why subscribe?

- Spend less time learning and more time coding with practical eBooks and Videos from over 4,000 industry professionals

- Improve your learning with Skill Plans built especially for you

- Get a free eBook or video every month

- Fully searchable for easy access to vital information

- Copy and paste, print, and bookmark content

Did you know that Packt offers eBook versions of every book published, with PDF and ePub files available? You can upgrade to the eBook version at packt.com and as a print book customer, you are entitled to a discount on the eBook copy. Get in touch with us at customercare@packtpub.com for more details.

At www.packt.com, you can also read a collection of free technical articles, sign up for a range of free newsletters, and receive exclusive discounts and offers on Packt books and eBooks.

Other Books You May Enjoy

If you enjoyed this book, you may be interested in these other books by Packt:

Hands-On Data Warehousing with Azure Data Factory

Christian Coté, Michelle Gutzait, Giuseppe Ciaburro

ISBN: 978-1-78913-762-0

- Understand the key components of an ETL solution using Azure Data Factory and Integration Services
- Design the architecture of a modern ETL hybrid solution
- Implement ETL solutions for both on-premises and Azure data
- Improve the performance and scalability of your ETL solution
- Gain thorough knowledge of new capabilities and features added to Azure Data Factory and Integration Services

Azure Data Engineering Cookbook

Ahmad Osama

ISBN: 978-1-80020-655-7

- Use Azure Blob storage for storing large amounts of unstructured data
- Perform CRUD operations on the Cosmos Table API
- Implement elastic pools and business continuity with Azure SQL Database
- Ingest and analyze data using Azure Synapse Analytics
- Develop Data Factory data flows to extract data from multiple sources
- Manage, maintain, and secure Azure Data Factory pipelines
- Process streaming data using Azure Stream Analytics and Data Explorer

Packt is searching for authors like you

If you're interested in becoming an author for Packt, please visit `authors.packtpub.com` and apply today. We have worked with thousands of developers and tech professionals, just like you, to help them share their insight with the global tech community. You can make a general application, apply for a specific hot topic that we are recruiting an author for, or submit your own idea.

Share Your Thoughts

Now you've finished *Data Modeling for Azure Data Services*, we'd love to hear your thoughts! Scan the QR code below to go straight to the Amazon review page for this book and share your feedback or leave a review on the site that you purchased it from.

`https://packt.link/r/1-801-07734-7`

Your review is important to us and the tech community and will help us make sure we're delivering excellent quality content.

Index

www.ingramcontent.com/pod-product-compliance
Lightning Source LLC
Chambersburg PA
CBHW081501050326
40690CB00015B/2882